Systems and Uses of Digital Sciences for
Knowledge Organization

Digital Tools and Uses Set

coordinated by
Imad Saleh

Volume 9

Systems and Uses of Digital Sciences for Knowledge Organization

Edited by

Sahbi Sidhom
Amira Kaddour

WILEY

First published 2022 in Great Britain and the United States by ISTE Ltd and John Wiley & Sons, Inc.

ISTE Ltd
27-37 St George's Road
London SW19 4EU
UK

www.iste.co.uk

John Wiley & Sons, Inc.
111 River Street
Hoboken, NJ 07030
USA

www.wiley.com

Any opinions, findings, and conclusions or recommendations expressed in this material are those of the author(s), contributor(s) or editor(s) and do not necessarily reflect the views of ISTE Group.

Library of Congress Control Number: 2022940925

British Library Cataloguing-in-Publication Data
A CIP record for this book is available from the British Library
ISBN 978-1-78630-773-6

Contents

Chapter 5. Social Networking Application, Connections Between Visual Communication Systems and Personal Information on the Web   97

Marilou KORDAHI

Chapter 6. A New Approach of Texts and Writing Normalization for Arabic Knowledge Organization   119

Hammou FADILI

Chapter 7. Ebola Epidemic in the Congo 2018–2019: How Does Twitter Permit the Monitoring of Rumors?

Marc TANTI

Chapter 8. From Human and Social Indexing to Automatic Indexing in the Era of Big Data and Open Data.

Nabil KHEMIRI and Sahbi SIDHOM

Chapter 9. Strategies for the Sustainable Use of Digital Technology by the AWI in the Management of Knowledge and Cultural Communication on the "Arab World" 165

Asma ABBASSI

Introduction

The OCTA International Multi-Conference on "Organization of Knowledge and Advanced Technologies" is a large-scale scientific event that brings together researchers and R&D professionals focusing on ideas and common actions in the organization of knowledge. The main objective of this book is to define collaborative strategies, use advanced technologies in multiple research fields and outline applications of knowledge organization for society and its cultural, educational, economic and industrial development.

Moreover, in a collaborative way in the OCTA event, the main conjuncture between scientific and professional communities is to initiate future innovative projects, in order to bring public and private institutions closer to tomorrow's technological challenges.

In February 2020, the scientific projects involved in the OCTA Multi-Conference edition were as follows:

– *SIIE* (https://siie2019.loria.fr/ & www.siie.fr) on "Information Systems and Economic Intelligence". The SIIE international conference aims to promote dialog between experts and researchers from both public and private sectors on fundamental and experimental knowledge of Information Systems and Economic Intelligence (SIIE). Its goal is to develop technologies related to economic intelligence (EI) in a risk environment. The dynamics of EI (i.e. Competitive Intelligence) depend on mastering the knowledge and skills needed to design the best strategies and to ensure that decision-makers make the right decisions.

Introduction written by Sahbi SIDHOM and Amira KADDOUR.

– *ISKO-Maghreb* (https://isko-maghreb2019.loria.fr/ and www.isko-maghreb.org) on "Digital Sciences: Impacts and Challenges on Knowledge Organization". The ISKO international scholarly society is devoted to the theory and practice of organization, including the objective of the ISKO-Maghreb and the ISKO chapter in Maghreb countries, which continues to contribute to our understanding of the factors that organize knowledge and the phenomena that affect the information society. The actions to be undertaken by the scholarly society ISKO-Maghreb will have to take into account socio-cultural, cognitive and economic aspects in the strategic management of knowledge. In relation to the knowledge society, knowledge must be seen in the context of its dynamics, content and scientific and technological interactions with academics, business and politics (actors and institutions).

– *CITED* (https://cited2019.loria.fr/) on "Advanced Technologies, Renewable Energies and Economic Development". The international symposium CITED aims to bring together the work on concerted and reflective research into the establishment of sustainable economic development based on technological advances, the optimal use of means and resources, and renewable energies. Joseph Aloïs Schumpeter (1934) highlighted the relationship between the innovation factor and the economic conjuncture: the pattern of economic transition, i.e. the theory of economic development. When analyzed in the context of economic cycles, the horizons of 2020–2030 appear, according to the cycles of Kondratiev (2014), as the beginning of the transition to a new era of production, industrialization and means, which can now be explained by the rise of the green economy.

– *TBMS* (https://tbms2019.loria.fr/) on "Big-Data-Analytics Technologies for Strategic Management: Innovation and Competitiveness". The International Symposium TBMS explores the practical implications of Big Data and how it reconfigures relationships, expertise, methods, concepts and academic knowledge in all sectors: social, professional and economic. Today, we have more data than ever before in human history. Data volumes multiplied by 100 between 1987 and 2007, and then doubled on average every year. This is an increase infinitely greater than that caused by the invention of printing (starting with J. Gutenberg), which had resulted in a doubling of data over 50 years. In an enthusiastically transdisciplinary way, Big Data orients us towards the classification of reality into categories generated by data, instead of imposing these classifications as input at the beginning.

Taking into account transdisciplinarity, OCTA asks the following questions:

– How can we strengthen alliances between multi-disciplinary and trans-disciplinary studies?

– How can we broaden our skills surrounding common objects of study?

– How can we innovate the solutions found and propose sustainable development to society confidently?

I.1. Scientific challenges

The selected proposals, which form the OCTA edition from February 2020, mainly represent the topic of "Systems, Tools and Digital Uses for Knowledge Organization". This scientific and multidisciplinary orientation wishes to point out the major concerns around the "Digital Uses for Knowledge Organization" with the proposals of "Models, Systems and Tools" to achieve this.

This book is a result of intensive and collaborative work between highly respected scientific authors. The nine chapters that have been selected for this book have been peer-reviewed by the OCTA program committee, both as written submissions and when presented during the OCTA multi-conference organization. In these circumstances of an exchange between authors and the scientific audience of OCTA, the proposals are enriched by authors at our request as chairs, and produce for this book an excellent reference in the "Information and Communication Sciences" and in a new scientific domain of "Digital Sciences: Impacts and Challenges on Knowledge Organization".

I.2. Structure of this book

Chapter 1, *Multi-Agent System and Ontology to Manage Ideas and Represent Knowledge: Creativity Challenge* by Pedro Chávez Barrios, Davy Monticolo and Sahbi Sidhom presents and develops the implementation of an intelligent system to support idea management. This is the result of a multi-agent system (MAS) used in a distributed system with heterogeneous information as ideas and knowledge, after the results about an ontology to describe the meaning of these ideas. The authors argue that the intelligent system assists the participants of the creativity workshop to manage their ideas, thereby proposing an ontology dedicated to these ideas. During the creative workshop, many creative activities and collaborative creative methods are used by roles immersed in this creativity workshop event where they share their knowledge. The collaboration of these roles is physically distant, their interactions might be synchronous or asynchronous, and the information of ideas is heterogeneous, in order to confirm that the process is distributed. These ideas are written in natural language by participants who have a role, and are heterogeneous since some of them are described by schema, text or scenario of use. This chapter presents the MAS and the associated ontology design.

In **Chapter 2**, *Comparative Study of Educational Process Construction Supported by an Intelligent Tutoring System* by Walid Bayounes, Inès Bayoudh Sâadi and Hénda Ben Ghézala, the motivation observed by the authors is the need for educational processes that can be constructed and adapted to the needs of learners, the preferences of tutors and the requirements of system administrators and system designers of the intelligent tutoring system (ITS). In this context, this chapter explores the theory of studying the problem of educational process construction. This study introduces a new multi-level view of educational processes. Based on the proposed view, a faceted definition framework conducts a comparative study on the ITS to understand and classify issues in educational process construction.

Chapter 3, *Multi-Criteria Decision-Making Recommender System Based on Users' Reviews* by Mariem Briki, Sabrine Ben Abdrabbah and Nahla Ben Amor, argues that due to the huge increase in the volume of available online data, finding relevant information becomes a very challenging task. Recommender systems (RS) achieve personalization by capturing users' interests and provide them with items that would probably match their expectations and tastes. The objective of this chapter is to use text mining techniques to analyze users' reviews and to detect the multi-faceted representation of the active user's interests. The authors assume that analysis of textual data can reveal additional hidden information, allowing the profiling of users. They propose a multi-criteria text mining-based RS (MCTMRS) that defines different features/criteria of the recommended item and then builds the corpus of information of each criterion. Finally, users' reviews are exploited to capture users' interests in different criteria and to detect the items that satisfy users' multi-criteria preferences. The authors tested this on a real database extracted from the TripAdvisor website. The experimental study shows that the proposed solution improves accuracy compared to traditional approaches.

Chapter 4, *Spammer Detection Relying on the Reviewers' Behaviors Features Under Uncertainty* by Malika Ben Khalifa, Zied Elouedi and Eric Lefèvre, reports that nowadays the success of different companies and organizations depends on their e-reputation. The latter is manipulated by online reviews that not only have a fundamental impact on the company's development but also deeply influence the buying decision of readers. That is why spammers post fake reviews to cheat online review systems, in order to mislead consumers and damage e-commerce. This makes spammer detection one of the most important tasks to stop fraudulent online activities and protect the e-reputation of companies, restaurants, hotels and brands. Hence, we put forward a novel approach that differentiates between spammers and innocent reviewers. The method presented by the authors is based on the K-nearest neighbor algorithm in the belief function theory and relies on suspicious behavior indicators which are considered as features. The experimental study shows the performance and robustness of the method, which was tested on two real-world labeled datasets extracted from yelp.com.

Chapter 5, *Social Networking Application, Connections Between Visual Communication Systems and Personal Information on the Web* by Marilou Kordahi, contributes to the field of information systems by examining the connections between social networking sites and visual communication systems. This chapter clearly presents the design of an innovative social networking application and develops a prototype of it: "SignaComm". It enables multilingual communication between users worldwide and in various situations (e.g. protection of personal information on the Web). SignaComm is based on the theory of patterns as well as the principles of ontologies and the signage system. The theory of patterns allows the reuse of patterns to serve as assets for programming advancement and critical thinking. Ontologies define structured concepts and objects by providing significance to an information system in a particular field, and allow the development of connections between these concepts and objects. The signage system is used to provide information on a topic, in order to facilitate the communication between users on an international scale. The "signagram" is its writing unit. When creating the SignaComm, we use a machine translation of key phrases into signagrams. After writing the prototype with programming languages such as Python, PHP and Javascript, we test its capabilities to communicate instant messages to users.

Chapter 6, *A New Approach of Texts and Writing Normalization for Arabic Knowledge Organization* by Hammou Fadili, argues that "knowledge organization" is an important research field, as evidenced, for example, by numerous research works on classification. This last discipline is complicated, especially in the case of Arabic data from the Web, because Arabic has many forms of writing, spelling, structure, etc. on the one hand, and the lack of preprocessed and normalized data, on the other. Implementing solutions that can help remedy these problems is a real need and a big challenge for the standardization process that this language must know, especially in the new world of publishing that is the Web. This is characterized by many forms of writing styles where everyone writes in their own way without any constraints. It is in this context that the author suggests a new approach based on unsupervised deep learning methods and implements a system to help the normalization of Arabic texts and writing, in order to facilitate and improve their organization and classification.

Chapter 7, *Ebola Epidemic in the Congo 2018–2019: How Does Twitter Permit the Monitoring of Rumors?* by Marc Tanti, presents the Ebola epidemic that has mainly affected three West African countries (Guinea, Sierra Leone and Liberia), killing 20,000 people. A number of articles have studied the rumors that circulated during this outbreak on Twitter. For example, Fung's article pointed out that false information about the treatment of the disease, such as bathing in salt water for healing, was disseminated on the platform (Fung 2016). Jin's article also pointed out that these media were behind a fake news article alleging that a snake was at the origin of the epidemic. This article also listed the top 10 rumors circulating on Twitter (Jin 2014).

This study reports that the Ebola virus has been raging in the Democratic Republic of the Congo (DRC) since August 1, 2018, killing more than 2,050 people. It is the second largest epidemic after West Africa. No studies have been conducted to determine who is communicating about this epidemic and what types of tweets or rumors are being disseminated.

To answer this question, the author conducted analysis on Twitter using the Radarly® software, over a period dating from April 1 to July 7, 2019. This work has highlighted several actors communicating around the epidemic (general public, experts, politicians, the media, etc.). This study found 12 main influencers and revealed a negative message rate (rumors) of 73.31%.

Chapter 8, *From Human and Social Indexing to Automatic Indexing in the Era of Big Data and Open Data* by Nabil Khemiri and Sahbi Sidhom, outlines that in the era of Big Data and Open Data, massive and heterogeneous collections of documents (from text to multimedia) are created, managed and stored electronically. To make these documents more usable, a set of processes, like human (i.e. manual) indexing, social indexing and automatic (i.e. machine/algorithmic) indexing, make it possible to create representations of documents by a set of metadata, descriptors (or terms) and social tags. These representations will make it easier to find information in a massive and scalable collection of documents from different sources (document databases, social networks, Open Data, etc.), in order to respond to user *information needs* (i.e. user requests). Numerous research studies have been carried out to put forward indexing approaches depending on the type of indexed documents, as well as to observe the evolution of indexing methods linked to the evolution of document representations, electronic content, Big Data and Open Data. This chapter presents a detailed overview, including a state of the art on approaches and methodologies from human/manual indexing, social indexing and automatic indexing with a set of algorithmic methods in the era of Big Data and Open Data.

Chapter 9, *Strategies for the Sustainable Use of Digital Technology by the AWI in the Management of Knowledge and Cultural Communication on the "Arab World"* by Asma Abbassi, shows that today more than ever, in the post-digital era, the management, production and transmission of content face major and multiple challenges. The stakes become all the greater when this content, which is seen, read and listened to, conditions knowledge and scholarship around a civilization or a geographical area. Such is the case with the Arab World Institute (AWI). Similarly, knowledge management and cultural communication have a special place in sustainable development, which has become a concept of primary interest in global policies for several years (CGLU 2015). The optimization of cultural tools has now become a necessity and even an emergency. This chapter lies in this context, as it is part of a several-year doctoral work focused on the study of the images of the "Arab world" produced by the AWI via its public activities. It is also based on a recent work that analyzes the digital communication strategy of the AWI and its digital media.

At the crossroads of digital humanities, cultural studies and information, communication sciences and sustainable development, the author suggests studying the relationships between the material and immaterial worlds and the management methods of knowledge through various digital tools, the images conveyed around the "Arab world" via the construction, organization and transmission of knowledge from the AWI. This study integrates the feedback issue and the degree of interaction of audiences, mainly those in the "Arab world", with a discussion on digital sustainability and durable knowledge management.

I.3. References

CGLU (2015). Culture 21 : Actions. Engagements sur le rôle de la culture dans les villes durables. *Cités et Gouvernements Locaux Unis*, Barcelona [Online]. Available at: https://www.arts-ville.org/wp-content/uploads/2017/10/Culture21 Actions_2015.pdf.

Fung, I.C.-H., Fu, K.-W., Chan, C.-H., Chan, B.S.B., Cheung, C.-N., Abraham, T., Tse, Z.T.H. (2016). Social media's initial reaction to information and misinformation on ebola, August 2014: Facts and rumors. *Public Health Reports*, 131(3), 461–473 [Online]. Available at: https://doi.org/10.1177/ 003335491613100312.

Jin, F., Wang, W., Zhao, L., Dougherty, E., Cao, Y., Lu, C.-T., Ramakrishnan, N. (2014). Misinformation propagation in the age of Twitter. *Computer*, 47(12), 90–94.

Kondratiev, N.D. (2014). *The Long Waves in Economic Life*. Martino Fine Books, Eastford.

Schumpeter, J.A. (1934). *The Theory of Economic Development: An Inquiry into Profits, Capital, Credit, Interest, and the Business Cycle*. Harvard University Press, Cambridge.

Multi-Agent System and Ontology to Manage Ideas and Represent Knowledge: Creativity Challenge

1.1. Introduction

Every year, The University of Lorraine organizes a creativity workshop called "48 hours generating ideas" (48H). We observed that thousands of ideas (IdC) were generated during the last 48H creativity workshop in 2017 (Ecole d'été RRI 2017). In order to manage these ideas, a multi-agent system is studied and proposed since it has been proved to be efficient in a distributed process and to propose an ontology to represent knowledge. The concept of MAS appeared at the end of the 20th century. The multi-agent system has two forms of vision: the interaction among agents and the interaction among humans, *the first, as an artificial intelligence (AI) concept attributed to Nils Nilsson "all AI is distributed-1980" and the second as artificial life (Alife) based in the complex adaptive behaviors of communities of humans* (Weyns et al. 2004). B*y relating an individual to a program, it is possible to simulate an artificial world populated with interacting processes* (Drogoul et al. 1992; Chávez Barrios 2019; Chávez Barrios et al. 2020). The individual is an agent who interacts according to their environment, which is clearly defined with respect to reality. These interactions among agents and their environments are an important aspect of the MAS. At the beginning of the 21st century, *an initial model tool was used to create generic multi-agent platforms based on an organizational mode based on the core model*

Chapter written by Pedro Chávez Barrios, Davy Monticolo and Sahbi Sidhom.

agent-role-group (Gutknecht and Ferber 2000), and also *multi-agent model is used involving some agents to hundreds focusing on the breakdown of a problem into simple ones that the agent can solve* (Simonin and Ferber 2003). At present, MAS has been used to improve energy efficiency (Zhang et al. 2014). However, four topics were covered during the creativity workshop 48H: a) our intelligent system based in agents is complicated due to the multiple and different interactions among actors and roles without our target losing the ability "to assist participants of the creativity workshop to manage their ideas" and to develop an annotation system; as Ferber said *"organization is the interaction of different groups"* (Ferber 1994); b) collaborative interactions among actors and roles that are distant must be defined and every role might represent an agent; c) divergence actions are used by participants to create ideas using creative activities and collaborative creative methods; and d) convergence actions are used to produce idea cards as well as to store, share, evaluate and improve them. Several design methodologies of multi-agent systems exist: GAIA (Wooldridge et al. 2002; Zambonelli et al. 2003) and DOCK (Girodon et al. 2015b) are examples of these designs. Multi-agent systems have two principal methodologies (Esparcia et al. 2011), *those oriented towards agents and those oriented towards organizations* are *based on organizational units, services, the environment and norms* (Argente et al. 2009). Due to the uncountable times that agents are mentioned and the interaction of agents in a multi-agent system, we have to present some definitions about the concept of the agent; it has some primordial functions for the creation of our intelligent system. Since the last century and the beginning of this century, the concept of the agent and its characteristics has appeared. *There are several definitions about agents, description of agents' requirements, uses of agents* (Shen and Norrie 1999) *and descriptions of agents' evolution* (Vanhée et al. 2013). In software design, an agent represents structured aggregations *of elements like events, actions, beliefs, plans and tasks* (Kinny 2002). In actual definition, "agent" is a system whose behavior is teleonomic (Castelfranchi 2012), "goal-oriented" towards a certain state of the world; in another definition, an agent is specified as *an active communicating entity which plays roles inside groups* (Gutknecht and Ferber 2000). *A final and, in our opinion, very important definition is "an agent is a computer system with independent action with respect to interventions of humans of other agents"* (Jennings and Wooldridge 1998).

Having put forward the MAS, the next step is to define ontology. It is dedicated to ideas and specifically to assist participants in idea generation during the creativity workshop. In addition, our ontology represents knowledge from this CWS like ideas, processes, activities, actors, roles, methods, idea cards and possible solutions; this ontology is used to annotate ideas and to facilitate idea management. As an initial definition of the word, the etymology of ontology comes from ancient Greek, but the concept of ontology appeared in the 17th century in the work "Ogdoas Scholastic" by Jacob Lorhard (Øhrstrøm et al. 2008). Also, in the last century, the concept of ontology focused on the definition of objects, concepts, entities and relationships among them in a defined area (Genesereth and Nilsson 1987; Gruber 1995), and ontology works as a database with information, properties and relationships about concepts that exist in the world or a particular domain (Mahesh 1996).

This chapter presents the results about our multi-agent and ontology proposals. Initially, we present the context, problem and methodology; section 1.2 is about the state of the art of MAS and ontology approaches, and section 1.3 describes our approaches. Finally, the last section is dedicated to results and conclusions.

1.2. Multi-agent system (MAS) and ontology

1.2.1. *MAS and ontology*

1.2.1.1. *Multi-agent system (MAS)*

Inside the MAS, the concept of the agent is vital. The definition of agent in a general and complete AI idea: *an agent could be human or robotic, the human agent acts through their physical sensors and acts with their extensions; the robot acts with its electronic sensor and effectors* (Russell and Norvig 2003).

There are different varieties of agents, but in a robotic sense, *autonomous agents should be reactive to changes in the environment and must be able to predict incompatible goal management and adaptivity (prediction)* (Decugis and Ferber 1998). An agent, in our environment of creativity workshop 48H, is part of a team and can play one or more roles. An agent has several properties, such as the capacity to interact with their environment, the capacity to communicate with other agents, the necessity to achieve an objective, the capacity to manage their resources, the capacity to perceive the environment, the capacity to represent the environment partially or totally and eventually the capacity of reproduction (Ferber 1995). In Wooldridge's conceptions, an agent is an informatic system in a specific environment, with autonomous actions to achieve its objectives (Wooldridge et al. 1999). A final definition is that agents interact amongst themselves to achieve a general objective, and there exist two kinds of agents, cognitive and reactive (Ferber 1995). Artificial intelligence (AI) is an important discipline that defines an agent in different ways. *AI borrows concepts (states, actions and rational agents) and techniques for autonomic computing* (Kephart and Walsh 2004).

1.2.1.2. Ontology

In the last century, the concept of ontology focused on the definition of objects, concepts, entities and relationships between them in a defined area (Genesereth and Nilsson 1987; Gruber 1995), and ontology works as a database with information, properties and relationships about concepts that exist in the world or a domain (Mahesh 1996).

Berners-Lee suggested using ontologies in the context of the Internet in order to bring a semantic dimension of the Web and to use collections of information called ontologies that is a component of the semantic web (Berners-Lee and Hendler 2002) that can be shared and reused (Elbassiti and Ajhoun 2014). The semantic web allows us to build ontologies by using a set of languages such as RDF, RDFs and OWL to structure knowledge resources.

The creation of a domain ontology needs to define the concepts, procedures, activities and relationships that belong specially to this domain or field in detail, trying to eliminate ambiguity and doubts due to the communication among web researches and machines (computers) using software applications, as explained by Staab and Studer (2004).

According to Mathieu d'Aquin, "ontologies represent the essential technology that enables and facilitates interoperability at the semantic level, providing a formal conceptualization of the data which can be shared, reused, and aligned" (D'Aquin and Noy 2012).

There are several existing libraries dedicated to ontology using some formats to link their information, such as RDF, which guarantees the interoperability, making it possible for applications to reuse data and to link diverse data (D'Aquin and Noy 2012). BioPortal is a library of biomedical ontologies developed by the National Center for Biomedical Ontologies (Noy et al. 2009).

1.2.2. MAS methodologies

There are some methodologies to design a multi-agent system (MAS). These methodologies involve mainly roles, agents, interactions among agents and the environment. The first methodologies were developed at the end of the century based on interaction between roles (Wooldridge et al. 1999), but in this century, several methodologies have appeared, such as ICTAM (Elsawah et al. 2015), DOCK (Girodon et al. 2015a), MOBMAS (Tran and Low 2008), ADELFE (Picard and Gleizes 2006) and GAIA (Wooldridge et al. 2000; Zambonelli et al. 2003). Here, we will describe the most important characteristics of each one.

1.2.2.1. Wooldridge's GAIA methodology

This technique uses analysis and design; it moves from abstract to gradually more concrete concepts; from analysis, roles and interaction models; from design agent, services and acquaintance models (Wooldridge et al. 2000). We used it in the 48H organization due to the details of the system and the full definition of agents.

1.2.2.2. DOCK methodology

DOCK's approach (Gabriel 2016) allows us to model MAS based on knowledge management. This methodology describes an intelligent knowledge system; it uses human organizations. The DOCK methodology defines four elements: the organizational structure to identify agents (roles) and also the process, the activity and the role models to define every simple piece of knowledge in detail. The model defines three stages: human organization, agent organization and interactions.

1.2.2.3. Other methodologies

There are other styles to analyze and design MAS MOBMAS uses ontology (Tran and Low 2008), which describes that most of the methodologies are deficient in design, organization and interaction of agents; for this reason, MOBMAS is focused mainly on five descriptive activities focused in agents. The SABPO style tries to achieve the goals of the organization as a main target of the agent's interactions (Petta et al. 2003).

The ADELFE, MOBMAS and SABPO techniques require agents as well as identification and definition of agents and their principal cooperative activities (Picard and Gleizes 2006; Rougemaille et al. 2009).

1.2.3. *Methodologies to design ontologies*

Ontology requires a well-defined process to represent reality. Several methodologies already exist to build ontologies. We will present some approaches to build ontologies.

a) Enterprise approach (Uschold and King 1995; Jones et al. 1998) is based on four steps: identification of the purpose, of the scope, of terms and finally formal evaluations; it might have a cycle from formal evaluations to identifying the scope.

b) Methontology: the goal of this methodology defined by Fernandez and Goméz-Pérez (Fernandez et al. 1997) is to clarify the activities they should perform. They claim ontology is an art and try to transform into engineering. It has seven steps: specification to obtain natural languages ontology, knowledge acquisition, conceptualization, integration to get uniformity, implementation, evaluation and documentation.

b) The KBSI IDEF5 is devoted to assisting in the creation, modification and maintenance of ontologies according to Jones in Jones et al. (1998) and Peraketh et al. (1994). The process has five steps: organizing and scope, data collection, data analysis, initial ontology and evaluation, as well as a final step to design ontologies,

d) Methodology to design ontologies from organizational models, the phases and activities applied to creativity workshops that represent the ontology process (Gabriel et al. 2019). This describes an ontology to model knowledge in a creativity workshop, its description:

Phase 1: Definition

– definition of domain, scope and purpose;

– definition of the question-skills of the ontology (aptitudes).

Phase 2: Conceptualization

– conceptualization and acquisition;

– reuse of existing ontology concepts.

Phase 3: Development

– development of the ontology (programming, formalization);

– population of the ontology.

Phase 4: Validation/Evaluation

– evaluation.

1.3. MAS and ontology: our approach proposal

The MAS requires the most detailed information about agents and the methods described here try to achieve this but GAIA has provided initial definitions since the beginning of this century. With respect to the ontology, every previous ontology process has roughly scope, purpose, concepts, relationships and evaluation dedicated to its domain; for this reason, we work with GAIA and the enterprise approach ontology.

1.3.1. *MAS methodology GAIA*

The models in GAIA have two phases, first the analysis phase with the role model and interaction model and the second phase, the design phase (see Figure 1.1).

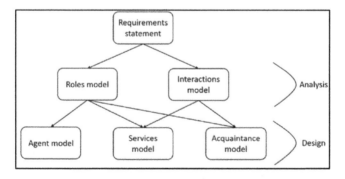

Figure 1.1. *Relationships between the GAIA models (Picard and Gleizes 2006, p. 3)*

The design processes of GAIA have three models, agent model, services model and acquaintance model that help us to understand the roles and interactions described before in the analysis phase (Wooldridge et al. 2000).

The objectives of a multi-agent system are to manage idea cards, to take decisions and to enhance creative techniques by means of the reactive and cognitive agents in the environment of the 48H creativity workshop. The most abstract concept is the system. The organization is a collection of roles and interactions among them (see Figure 1.2).

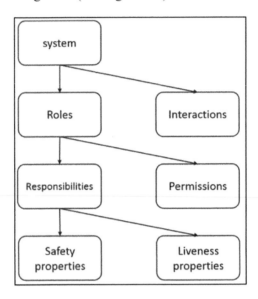

Figure 1.2. *Analysis of concepts (Wooldridge et al. 2000, p. 4)*

1.3.2. *Applying the ontology, Uschold's ontology*

The proposed methodology to build our ontology must help to represent the evolution of the ideas, the individual ideas, after-idea cards and finally the possible solutions, coming out of this organizational model called 48H. This methodology follows a process based primordially on building the ontology. Initially, it expresses the meaning of the organization after it works on the design of ontology, finally the judgment and documentation.

The methodology chosen is the enterprise ontology of Uschold (Uschold and King 1995). It has four phases. With this ontology, we can easily define the phases and identify the steps with the finality to do several iterations to correct our job. The Uschold ontology (see Figure 1.3) focuses mainly on building the ontology that is something that we appreciate; for us, the capture, coding and integrating existing ontologies are vital steps, without forgetting the iteration to improve.

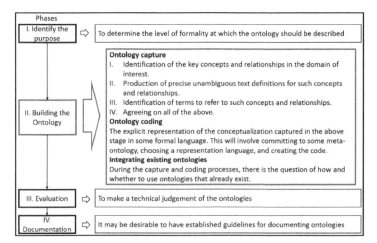

Figure 1.3. *Ontology Uschold phases. For a color version of this figure, see www.iste.co.uk/sidhom/systems.zip*

1.4. Results

1.4.1. *Multi-agent system results*

1.4.1.1. *Analysis: Roles model*

We will describe the roles of the agents, and we have chosen to give the same role to the agents as to the participants of the creativity workshop, i.e. "creative expert", "industrial manager", "organizer", "solver participant" and "technical expert". The objective of the agents is to assist the participants to achieve their activities and to manage their ideas during the creativity workshop process.

The schema of the solver participant (see Table 1.1) details the production of ideas in idea cards. There are 15 protocols (shown in the row Protocols and activities in Table 1.1) where the solver participant acts. The permissions produce an individual idea using individual activities, to produce at least two idea cards using a collaborative creative method, to have the same problem in WorkIdeaCards at the time of sharing with a colleague team and to evaluate idea cards from the same problem except our idea card.

Role schema	Name
Description and objective	The role solver participant: To produce ideas in an individual way using activities. To produce idea cards by means of collaborative creative methods.
Protocols and activities	RequirementsInscription (Name, Last name, Institution), GiveRequirements (Name, Last name, Institution), Assignation (Assign_InstToWork, Assigned_ind, assigned_rol), Provides (part_team, problem). Offer_activity, SelectActivity, WorkIdeas, Offer_method, SelectMethod, WorkIdeaCards, Improve, CompareIdeas, SendingIdeaCards, ReceivingPossibleSolutions, WatchingPossibleSolutions, AwardsEnd
Permissions	The actor must be registered as a solver participant; To produce at least one individual idea using individual activities. To produce two idea cards using a collaborative creative method. To have the same problem in WorkIdeasCards at the time of sharing with colleague team. To evaluate idea cards from the same problem except our idea cards and the idea cards from our team partner.
Responsibilities	
Liveness	Solverparticipant = (RequirementsInscription.GiveRequirements)+ · (Assignation)+ · (Provides)+ · (Offer_activity.SelectActivity.WorkIdeas)+ · (Offer_method.SelectMethod.WorkIdeaCards.Improve)+ ·(CompareIdeas)+ · (SendingIdeaCards.ReceivingPossibleSolutions)+ · (WatchingPossibleSolutions.AwardsEnd)+
Safety	Idea > 0 Idea card = 2 by team.

Table 1.1. *Role solver participant*

1.4.1.2. *Analysis: Interaction model*

This second model describes the communication protocols for each agent. The agent's protocol has some elements (see Figure 1.4) that help us to improve the explanation about the protocol's description (Colleman et al. 1996) in the interaction of the agents.

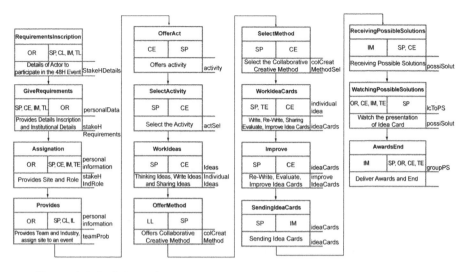

Figure 1.4. *Definition of protocols associated with the role solver participant. For a color version of this figure, see www.iste.co.uk/sidhom/systems.zip*

1.4.1.3. *Design: Agent model*

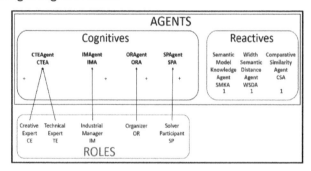

Figure 1.5. *Definition of protocols associated with the role solver participant. For a color version of this figure, see www.iste.co.uk/sidhom/systems.zip*

In the agent model (see Figure 1.5), we identify seven agents and their instances that will make up the system. During the creativity workshop, the five roles were easily identified according to the agent model proposed. The role creative expert and technical expert will form an agent called creative technological expert agent (CTEAgent-CTEA); the number of CTEA agents is one or more. The rest of the roles have their agents, noting that the agent organizer (ORAgent-ORA) has one or more instances. However, we add three agents, semantic model knowledge agent (SMKA), width semantic distance agent (WSDA) and the comparative similarity agent (CSA). These agents help us to order ideas according to semantic distance, width density and comparative similarity of idea cards.

1.4.1.4. *Design: Service model*

The principal services during the creativity workshop 48H are:

"To obtain information of actors and assignation of roles", "selection and application of activity for ideas", "selection and application of methods for idea cards", "evaluation by partners and improving idea cards as a goal", "classification of idea cards" and "sending possible solutions". These services (see Table 1.2) are functions that the agents have to execute according to the protocols described before.

Service	Inputs	Outputs	Pre-conditions	Post-conditions
Obtain information of actors and assignation of roles	Actor details Name, last name, institution, sex, date of birth	Actor requirements	Event = 1	Institution >= 1 Industry >= 1 Role >= 2 Team >= 2 Problem >= 1
Selection and application of activity for ideas	Group, creative technical expert, activities	Ideas	Ideas per participant at least in mind	Idea > 0
Selection and application of methods for idea cards	Thousands of ideas, many methods	Idea cards	2 idea cards per group = 2	Idea cards > 2
Evaluation by partners and improving idea card as a goal	Two ideas per group	Idea cards	2 idea cards per group	Idea cards > 2
Classification of idea cards	n idea cards	n idea cards	At least 2 idea cards	Idea cards > n
Sending possible solutions	Idea cards	Possible solutions	2 possible solutions per group	Possible solution >= 2

Table 1.2. *Role solver participant*

1.4.1.5. *Design: Acquaintance model*

The acquaintance model (see Figure 1.6) determines communication between the different agents. The agents creative expert CTEA, Organizational ORA, solver participant SPA and industrial manager agent IMA communicate during the entire creativity workshop. The agents CTEA and SPA also interact with the agents: semantic model knowledge SMKA and width semantic distance WSDA.

The agents SMKA, WSDA and CSA take action with the purpose of classifying the idea cards at the end of the creation, sharing, evaluation (among the partner group and the rest of the groups) and improving.

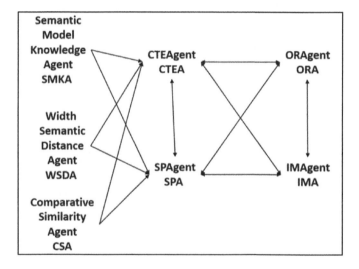

Figure 1.6. *Acquaintance model*

1.4.2. **Ontology results**

The domain is an ontology dedicated to creativity, and we called it "the collaborative creative ideas ontology", or CCIDEAS.

It aims to describe and inform all the concepts related to the creativity workshop challenge. These concepts are identified inside of an organizational model.

The purposes of this ontology are to represent knowledge and ideas and to understand the creativity workshop. The ontology will be used to compare ideas in the intelligent system.

1.4.2.1. *Creation of concepts*

These concepts and relationships determine the ontology. The definition of every concept that is part of this creativity workshop (see Table 1.3) will be used to create a triplet model based on three inter-related concepts.

Name of the concept	Definition
1 Activity	The action(s) that the actor follows to produce ideas (I) in the phase of divergence. The activities of divergence used to produce individual ideas. Type: String.
2 Actor	The person who will participate in the event and can take a role in solving problems. The concept will indicate the role or several roles to assume during a CWS. Type: String.
3 Collaborative creative method (CCM)	Set of instructions applied by solver participants to generate idea cards (IC). The methods of convergence are used to produce idea cards. Type: String.
4 Event	The name of the CWS and its edition. Type: String.
5 Idea	The individual ideas (I) are produced in the phase of divergence; they are created using combinational, exploratory and transformational techniques with individual activities. The actor captures the initial individual idea. Type: String.
6 IdeaDesc	Idea's description. Type: String.
7 IdeaCard	The result of the use of ideas (I) and collaborative creative methods (CCM). An actor with the role of the solver participant and from a team creates idea cards using CCM.
8 ICDesc	Idea card's description. Type: String.
9 ICTitle	Idea card's title. Type: String.
10 ICScenery	Idea card's scenery. Type: String.
11 ICPrioCli	Idea card's priority client. Type: String.
12 ICAdvant	Idea card's advantage. Type: String.

13 ICRisk	Idea card's risk. Type: String.
14 Industry	The name of the industry; this concept has the problem. The industrial manager proposes the industry that contains the problem. Type: String.
15 Problem	The reasons why the organization creates the creativity workshop CWS events every year. Industries like Assystem, Bostik, CEA Tech, Decathlon, GRDF, ICM, MSA Safety, Muller, Normande Aerospace, Pierre Fabre and Scarabée Biocop participate in those events. The problem is assigned to a team by means of the industry. Type: String.
16 Role	The type of character that the actor takes (organizer, solver participant, creative expert, technical expert, Industrial manager). Type: String.
17 Site	The place where actors will work (ASU BAHRAIN, CESI NANTERRE, ENSGSI NANCY, etc.). The site given to an actor. Type: String.
18 Team	The set of actors with the same role (solver participant), event, problem, site, name of team and color. (Str_Ass_1, Lyo_Assy_1, Uca1, Str_Ass_2, etc.). An actor takes part of a team. Type: String.
19 Vocabulary	The vocabulary is formed by ADJECTIVE, ADVERB, NOUN, VERB, ARTICLE, PRONOUN, PREPOSITION, CONJUNCTION AND INTERJECTION; some fields of the idea card's concept use these concepts such as title, description, priority client, name, scenery, advantages and risk. The vocabulary is part of idea's description. Type: String.
20 Organizer	The actor who takes the role in activities of assignation of roles, industries, team and event. Initially, this role creates the event and asks for information from the actors with the purpose of doing the inscription. Type: String.
21 Solver participant	This concept is part of the roles and part of the team. The site is assigned to the solver participant; other relationships are: Select activity, Create ideas, Send possible solutions. Type: String.
22 Creative expert	The creative expert offers activities and collaborative creative methods to the team of solver participants. Type: String.
23 Technical expert	Technical expert helps teams to improve idea cards. Type: String.
24 Industrial manager	The industrial manager proposes an industry to the creativity workshop. Type: String.
25 Possible Solutions	The concept possible solutions are the idea cards with best score according to the semantic skills like width, semantic distance and similarity.

Table 1.3. *Definition of concepts*

1.4.2.2. *Creation of relationship proposed*

The creation of relationships uses the format subject–verb–object with the purpose of creating sets of three elements:

– the term of relationships;

– the concepts (Bachimont 2000) where the relationship represents the verb (see Table 1.4);

– and present the global ontology (Figure 1.7).

1.5. Conclusion

Our main contribution to MAS and ontology is to assist participants during the creativity workshop to manage their ideas and knowledge and to put forward an intelligent system based in MAS and to develop an annotation system (ontology). The global frontiers of the system (see Figure 1.8) show the relationship between the agents SKMA, WSDA and CSA with the objects of the system. The creativity support system interface human machine (see Figure 1.9) presents the agents, events and operations (Chávez Barrios 2019; Chávez Barrios et al. 2020).

With respect to agents: the agents inside our MAS manage ideas, but in the future, other agents could be focused to assist participants in the selection of a better activity or collaborative creative method. With respect to ontology: the ontology system (see Figure 1.7) presents the quantity of 25 concepts and 34 relationships; the heart of the ontology is formed by the idea card, actor and method.

1.6. Appendices

Relation name proposed	Domain (concepts)	Range (concepts)	Triplet and/or definitions
1 Select	Solver participant	Activity. Examples: brainstorming, write storming, bend it and shape it, brain borrow, copy cat, etc.	Solver participant selects activity. The property indicates that the solver participant selects an activity to create individual ideas during the phase of divergence.
2 Offers	Creative expert	Activity. Examples: brainstorming, write storming, bend it and shape it, brain borrow, copy cat, etc.	Creative expert offers activity. The property indicates that creative offers an activity.

3 Plays	Actor	Role. Examples: creative expert, technical expert, industrial manager, solver participant and organizer	Actor plays a role.
4 Assign	Organizer	Role. Examples: creative expert, technical expert, industrial manager and solver participant	Organizer assigns a role. The property indicates that the organizer assigns all the roles in the creativity workshop.
5 Propose	Industrial manager	Industry. Examples: Decathlon, ICM, Bostik, etc.	Industrial manager proposes an industry. The property indicates that the industrial manager proposes an industry.
6 Create	Organizer	Event. Examples: 48h InnovENT-Edition 2016, Operation 2015 InnovENT-E 48 hours to bring ideas to life	Organizer creates an event.
7 Assign	Organizer	Site. Examples: INSA LYON, ENSGSI, UCA MARRAKECH, etc.	Organizer assigns site.
8 Assign	Organizer	Industry. Examples: Decathlon, ICM, Bostik, etc.	Organizer assigns industry.
9 Assign	Organizer	Team. Examples: Nan_Dec_1, Nan_Dec2, Str_Ass_2, etc.	Organizer assigns teams.
10 Requires	Organizer	Actor. Examples: any institutional, educative or industrial person interested in creativity and solving problems	Organizer requires actor.
11 Help	Technical expert	Team. Examples: Nan_Dec_1, Nan_Dec2, Str_Ass_2, etc.	Technical expert helps team. The property indicates that technical experts help teams.
12 IsAssignedTo	Industry	Team. Examples: Nan_Dec_1, Nan_Dec2, Str_Ass_2, etc.	Industry is assigned to team.
13 Receive	Industrial manager	Possible solutions	Industrial manager receives possible solutions. The property indicates that industrial manager receives the possible solutions.
14 IsPartOf	Actor	Team. Examples: Nan_Dec_1, Nan_Dec2, Str_Ass_2, etc.	Actor is part of team.

15 IsAssignedTo	Site	Event. Examples: 48h InnovENT-Edition 2016, Operation 2015 InnovENT-E 48 hours to bring ideas to life	Site is assigned to an event.
16 Send	Solver participant	Possible solutions	Solver participant sends possible solutions. The property indicates that solver participant sends the possible solutions.
17 Present	Team	Possible solutions	Team presents possible solutions. The property indicates that team presents the possible solutions.
18 IsAssignedTo	Site	Role. Range: technical expert, solver participant and creative expert	Site is assigned to role.
19 IsAssignedTo	Team	Role. Range: technical expert, solver participant and creative expert	Team is assigned to role.
20 Create	Team	Idea card	Team creates idea card. The property indicates that team creates the idea cards.
21 Improve	Team	Idea card	Team improves idea card. The property indicates that team improves the idea cards.
22 Select	Team	CCM. Examples: six hats of thinking, the shirt off your back, puzzle pieces, organizational brainstorms, best off, rice storm, etc.	Team select CCM. The property indicates that team selects the collaborative creative method.
23 Use	CCM	Idea	CCM uses ideas.
24 Form	Idea	Idea cards	Ideas form idea card.
25 Use	Idea card	CCM. Examples: six hats of thinking, the shirt off your back, puzzle pieces, organizational brainstorms, nest off, rice storm, etc.	Idea card uses CCM.
26 Offer	Creative expert	CCM. Examples: Six hats of thinking, the shirt off your back, puzzle pieces, organizational brainstorms, best off, rice storm, etc.	Creative expert offers CCM.

27 IsPartOf	IdeaDesc	Idea	IdeaDesc is part of idea. The property indicates that idea description (IdeaDesc) is part of the idea.
28 Create	Solver participant	Idea	Solver participant creates idea. The property indicates that solver participant creates ideas.
29 IsPartOf1	ICDesc	Idea card	ICDesc is part 1 of idea card. The property indicates that the field idea card description (ICDesc) is part of the idea card.
30 IsPartOf2	ICTitle	Idea card	ICTitle is part 2 of idea card. The property indicates that the field idea card title (ICTitle) is part of the idea card.
31 IsPartOf3	ICScenery	Idea card	ICScenery is part 3 of idea card. The property indicates that the field idea card scenery (ICScenery) is part of the idea card.
32 IsPartOf4	ICPrioCli	Idea card	ICPrioCli is part 4 of idea card. The property indicates that the field idea card priority clients (ICPrioCli) is part of the idea card.
33 IsPartOf5	ICAdvant	Idea card	ICAdvant is part 5 of idea card. The property indicates that the field idea card advantage (ICAdvant) is part of the idea card.
34 IsPartOf6	ICRisk	Idea card	ICRisk is part 6 of idea card. The property indicates that the field idea card risk (ICRisk) is part of the idea card.

Table 1.4. *Definition of relationships*

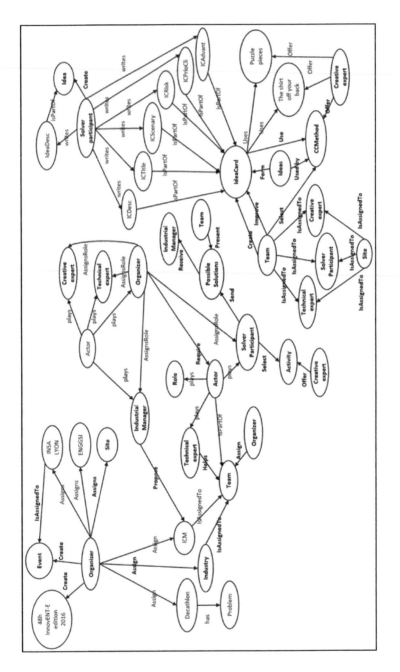

Figure 1.7. *Global ontology*

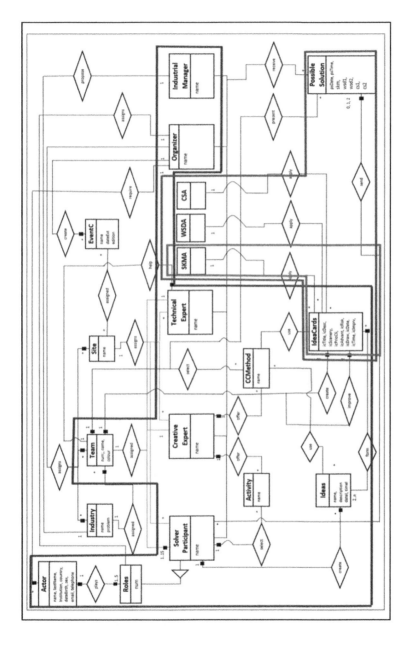

Figure 1.8. *Global frontiers of the system. For a color version of this figure, see www.iste.co.uk/sidhom/systems.zip*

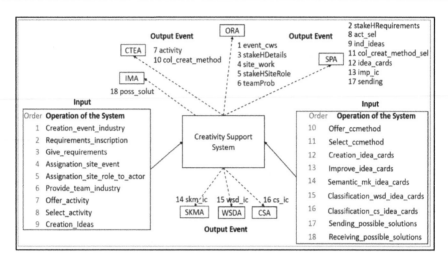

Figure 1.9. *Creativity support system interface human machine. For a color version of this figure, see www.iste.co.uk/sidhom/systems.zip*

1.7. References

Argente, E., Julian, V., Botti, V. (2009). MAS modeling based on organizations. *Agent-Oriented Softw. Eng. IX 9th Int. Work.*, 16–30.

Bachimont, B. (2000). Engagement sémantique et engagement ontologique : conception et réalisation d'ontologies en ingénierie des connaissances. *Ingénierie des connaissances : évolutions récentes nouveaux défis*, 505.

Berners-Lee, T. and Hendler, J. (2002). The semantic web. *Sci. Am.*, 21 [Online]. Available at: http://www.w3.org/2002/07/swint.

Castelfranchi, C. (2012). Guarantees for autonomy in cognitive agent architecture, *Intelligent Agents, ECAI-94 Workshop on Agent Theories, Architectures, and Languages*. Amsterdam, August 8–9.

Chávez Barrios, P. (2019). Design of information and technology tools to support the process of creativity and innovation. PhD Thesis, Université de Lorraine.

Chávez Barrios, P., Montocolo, D., Sidhom, S. (2020). Results of multi-agent system and ontology to manage ideas and represent knowledge in a challenge of creativity. *OCTA Multi-conference Proceedings: Information Systems and Economic Intelligence (SIIE)*, 208, February, Tunis, Tunisia [Online]. Available at: https://multiconference-octa.loria.fr/multiconference-program/.

Coleman, D. and Arnold, P. (1996). Fusion, La méthode orientée-object de 2e génération. *Méthodologies du logiciel*, ISSN 1242-5656

D'Aquin, M. and Noy, N.F. (2012). Where to publish and find ontologies? A survey of ontology libraries. *J. Web Semant.*, 11, 96–111.

Decugis, V. and Ferber, J. (1998). Action selection in an autonomous agent with a hierarchical distributed reactive planning architecture. *AGENTS '98 Proc. Second Int. Conf. Auton. Agents*, 354–361.

Drogoul, A., Ferber, J., Cambier, C. (1992). Multi-agent simulation as a tool for analysing emergent processes in societies [Online]. Available at: https://horizon.documentation.ird.fr/exl-doc/pleins_textes/pleins_textes_5/b_fdi_31-32/35495.pdf.

Ecole d'été RRI (2017). Les 28–29 août 2017 : "L'innovation agile : quels défis pour les individus, les organisations et les territoires ?" [Online]. Available: https://rni2017.event.univ-l.

Elbassiti, L. and Ajhoun, R. (2014). Semantic representation of innovation, generic ontology for idea management. *J. Adv. Manag. Sci.*, 2(1), 128–134.

Elsawah, S., Guillaume, J.H.A., Filatova, T., Rook, J., Jakeman, A.J. (2015). A methodology for eliciting, representing, and analysing stakeholder knowledge for decision making on complex socio-ecological systems: From cognitive maps to agent-based models. *J. Environ. Manage.*, 151, 500–516.

Esparcia, S., Argente, E., Botti, V. (2011). An agent-oriented software engineering methodology to develop adaptive virtual organizations. *IJCAI Int. Jt. Conf. Artif. Intell.*, (1), 2796–2797.

Ferber, J. (1994). Coopération réactive et émergence. *Intellectica*, 19, 19–52 [Online]. Available at: https://www.persee.fr/doc/intel_0769-4113_1994_num_19_2_1460.

Ferber, J. (1995). *Les systèmes multi-agents vers une intelligence collective.* InterEditions, Paris.

Fernández-López, M., Gómez-Pérez, A., Juristo, N. (1997). METHONTOLOGY: From ontological art towards ontological engineering. *AAAI-97 Spring Symposium Series*, Stanford University, EEUU, 24–26 March.

Gabriel, A. (2016). Gestion des connaissances lors d'un processus collaboratif de créativité. Université de Lorraine, ERPI Lab.

Gabriel, A., Chávez Barrios, P., Monticolo, D. (2019). Methodology to design ontologies from organizational models: Application to creativity workshops. *Knowl. Eng. Manag. Appl. to Innov.*, 33(2), 148–159.

Genesereth, M. and Nilsson, N. (1987). Logical foundations of artificial intelligence: Nomotonic reasoning. *Logical Foundations of Artificial Intelligence* [Online]. Available at: https://www.sciencedirect.com/book/9780934613316/logical-found ations-of-artificial-intelligence.

Girodon, J., Monticolo, D., Bonjour, E. (2015a). How to design a multi agent system dedicated to knowledge management; the DOCK approach. *Computer Science*, 113–121.

Girodon, J., Monticolo, D., Bonjour, E., Perrier, M. (2015b). An organizational approach to designing an intelligent knowledge-based system: Application to the decision-making process in design projects. *Adv. Eng. Informatics*, 29(3), 696–713.

Gruber, T.R. (1995). Toward principles for the design of ontologies. *Int. J. Hum.-Comput. Stud.*, 43(5), 907–928.

Gutknecht, O. and Ferber, J. (2000). MadKit: A generic multi-agent platform. *Proc. Fourth Int. Conf. Auton.*, Agents – AGENTS, 78–79.

Jennings, N.R. and Wooldridge, M. (1998). *Applications of Intelligent Agents. Agent Technology: Formations, Applications and Markets.* Springer-Verlag, Berlin.

Jones, D., Bench-Capon, T., Visser, P. (1998). Methodologies for ontology development. *Computer Science*, June.

Kephart, J.O. and Walsh, W.E. (2004). An artificial intelligence perspective on autonomic computing policies. *Proc. – Fifth IEEE Int. Work. Policies Distrib. Syst. Networks*, 3–12.

Kinny, D. (2002). A visual programming language for plan execution systems. *Proc. First Int. Jt. Conf. Auton. Agents Multi-Agent Syst. (AAMAS-2002, Featur. 6th AGENTS, 5th ICMAS 9th ATAL)*, Bologna, 15–19 July.

Mahesh, K. (1996). Ontology development for machine translation: Ideology and methodology. Report, Comput. Res. Lab. New Mex. State Univ. MCCS-96-292.

Noy, N.F., Shah, N.H., Whetzel, P.L., Dai, B., Dorf, M., Griffith, N., Jonquet, C., Rubin, D.L., Storey, M.-A., Chute, C.G., Musen, M.A. (2009). BioPortal: Ontologies and integrated data resources at the click of a mouse. *Nucleic Acids Res.*, 37, 170–173.

Øhrstrøm, P., Shârphe, H., Uckelman, S. (2008). Jacob Lorhard's ontology: A 17th century hypertext on the reality and temporality of the world of intelligibles. In *Conceptual Structures: Knowledge, Visualization and Reasoning*, Eklund, P. and Haemmerlé, O. (eds). Springer, Berlin, Heidelberg.

Peraketh, B., Menzel, C., Mayer, R.J., Fillion, F., Futrell, M.T., DeWitte, P.S., Lingineni, M. (1994). Ontology capture method. Technical paper, KBSI, College Station.

Petta, P., Tolksdorf, R., Zambonelli, F. (2003). *Engineering Societies in the Agents World III*. Springer, Berlin, Heidelberg.

Picard, G. and Gleizes, M. (2006). The Adelfe methodology. In *Methodologies and Software Engineering for Agent Systems – The Agent-Oriented Software Engineering Handbook*, Bergenti, F., Gleizes, M.-P., Zambonelli, F. (eds). Springer, New York.

Rougemaille, S., Arcangeli, J.P., Gleizes, M.P., Migeon, F. (2009). ADELFE design, AMAS-ML in action: A case study. *Lect. Notes Comput. Sci. (including Subser. Lect. Notes Artif. Intell. Lect. Notes Bioinformatics)*, 5485 LNAI.

Russell, S. and Norvig, P. (2003). *Artificial Intelligence: A Modern Approach*, 2nd edition. Prentice Hall, Upper Saddle River.

Shen, W. and Norrie, D.H. (1999). Agent-based systems for intelligent manufacturing: A state-of-the-art survey. *Knowl. Inf. Syst.*, 1(2), 129–156.

Simonin, O. and Ferber, J. (2003). Un modèle multi-agent de résolution collective de problèmes situés multi-échelles. *RSTI/hors série, JFSMA*, 1–13.

Staab, S. and Studer, R. (2004). *Handbook on Ontologies*. Springer-Verlag, Berlin, Heidelberg.

Tran, Q.N.N. and Low, G. (2008). MOBMAS: A methodology for ontology-based multi-agent systems development. *Inf. Softw. Technol.*, 50(7–8), 697–722.

Uschold, M. and King, M. (1995). Towards a methodology for building ontologies. *Methodology*, 80(July), 275–280.

Vanhée, L., Ferber, J., Dignum, F. (2013). Agent-based evolving societies (extended abstract) categories and subject descriptors. *AAMAS '13: Proceedings of the 2013 International Conference on Autonomous Agents and Multi-Agent Systems*, May, 1241–1242.

Weyns, D., Parurak, V.D.H., Michel, F. (2004). Environments for multi-agent systems. *The Knowledge Engineering Review*, 00:0, 1–15.

Wooldridge, M., Jennings, N.R., Kinny, D. (1999). A methodology for agent-oriented analysis and design: A conceptual framework. *Proc. Third Int. Conf. Auton. Agents*, 69–76.

Wooldridge, M., Jennings, N.R., Kinny, D. (2000). The Gaia methodology for agent-oriented analysis and design. *J. Auton. Agents Multi-Agent Syst.*, 3(3), 285–312.

Wooldridge, M., Fisher, M., Huget, M., Parsons, S. (2002). Model checking multi-agent systems with MABLE. *AAMAS '02: Proceedings of the First International Joint Conference on Autonomous Agents and Multiagent Systems*, July, 952–959.

Zambonelli, F., Jennings, N.R., Wooldridge, M. (2003). Developing multiagent systems: The Gaia methodology. *ACM Trans. Softw. Eng. Methodol.*, 12(3), 317–370.

Zhang, W., Liu, W., Wang, X., Liu, L., Ferrese, F. (2014). Distributed multiple agent system based online optimal reactive power control for smart grids. *IEEE Trans. Smart Grid*, 5(5), 2421–2431.

Comparative Study of Educational Process Construction Supported by an Intelligent Tutoring System

2.1. Introduction

One of the evolving areas that will certainly occupy computer scientists over the next decade is the computer-supported learning environment. The latter is increasingly mediated by new technologies of information and communication. In fact, it can provide various kinds of technological support to guide teaching and learning (Rathore and Arjaria 2020). This context is therefore considered to be multidisciplinary as it includes computer science, cognitive psychology, pedagogy, didactics and educational sciences.

Throughout the years, there have been different types of environments, including the traditional e-learning system, instructional design system and intelligent tutoring system (Tzanova 2020). The latter is increasingly gaining popularity in the academic community due to its multiple learning benefits. In fact, most of these systems allow only the application of content adaptation and ignore the adaptation of educational processes (learning process and pedagogical process) (Bayounes et al. 2019). This is a major constraint for providing personalized learning paths and appropriate learning content and for exploiting the richness of individual differences in learning needs and pedagogical preferences (Bayounes et al. 2014). Therefore, this

Chapter written by Walid BAYOUNES, Inès BAYOUDH SÂADI and Hénda BEN GHÉZALA.

work more particularly concerns process adaptation in intelligent tutoring systems. In reality, personal learning paths are the best solution because students can more effectively acquire and retain knowledge and skills that will help them in the real world (Olesea 2019).

In fact, the main research problem is the construction of the learning process in ITS. The study of this problem leads to the extension of the construction in order to take into consideration the teaching process, also called the pedagogical process, which is strongly correlated with it. In the rest of this chapter, the term "educational processes" will be used to indicate both learning and pedagogical processes.

Within this context, a literature survey has been undertaken using a framework of educational process construction. The comparison framework is used in many engineering works in the literature and has proven its effectiveness in improving the understanding of various engineering disciplines (method engineering, process engineering) (Rolland 1998). By adopting the multi-level view of educational process definition, the goal of the proposed framework is to identify the suitable construction approach supported by an ITS in order to satisfy the individual learning needs and to respect the system constraints.

This research aims at providing an appropriate educational process that fits the needs of the learners, the preferences of the tutors and the requirements of the administrators of the intelligent tutoring system (ITS). The chapter thus starts by exploring the theory in order to study the problem of educational process construction. This study proposes a new multi-level view of educational processes. Based on this view, a faceted definition framework conducts a comparative study between different ITS to understand and classify issues in educational process construction. Finally, section 2.5 concludes this work with our contribution and research perspectives.

2.2. New view of educational process

The most important task of this research is to study the different definitions of educational processes. In fact, a valid process definition ensures valid process modeling to support an appropriate process adaptation (Chang et al. 2020).

The analysis of this study highlights the specification of multiple new levels of educational process definition (see Figure 2.1). These levels are divided into the psycho-pedagogical level, the didactic level, the situational level and the online level (Bayounes et al. 2020). The first two levels define the educational process in the theoretical layer. The last two levels define the practical layer. Both layers consider the correlation between pedagogical and learning facets.

This multi-level view specifies the guidance/alignment relationships between the learning and pedagogical facets and the instantiation/support relationships between various levels.

The different definition levels are specified by three major components (affective, cognitive and metacognitive). The first component defines the affective objective achieved by the educational process (why?). The second component defines the cognitive product adopted by the process (what?). The third component includes the metacognitive process that is used to achieve the affective objective (how?).

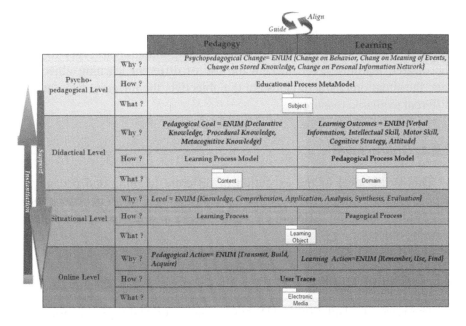

Figure 2.1. *Multi-level view of educational processes (Bayounes et al. 2020). For a color version of this figure, see www.iste.co.uk/sidhom/systems.zip*

2.2.1. *Psycho-pedagogical level*

At this level, the pedagogical and the learning view of the process are not defined. The educational process is viewed as a psycho-pedagogical change that occurs by using the educational process meta model. The latter adopts various subjects, which are defined by different theories of learning paradigm.

2.2.2. *Didactic level*

At this level, the process is specified by adopting the interaction between the pedagogical and the learning view. By considering the latter, the process model is used to achieve learning outcomes by considering the constraints of the learning domain. For the pedagogical view, the process is viewed as a pedagogical goal achieved via defining a process model based on pedagogical content.

2.2.3. *Situational level*

This is the level of the instantiation of process models by considering the different characteristics of the learning/teaching situation in order to reach the desired learning level through the use of different learning objects. In fact, the situation characteristics are the objective, the different tasks and the different available resources.

2.2.4. *Online level*

This is the level of the execution of different learning and pedagogical actions that are supported by different learning systems. This execution is achieved by adopting different electronic media in order to use the different learning objects.

2.3. Definition framework

The issue of adaptation in ITS has become a crucial topic of research in recent years. With the emergence of ITS, it has become possible to provide a learning process that matches the learner characteristics, the pedagogical preferences and the specific learning needs. In this regard, a literature survey has been conducted by using a framework of educational process construction (see Figure 2.2).

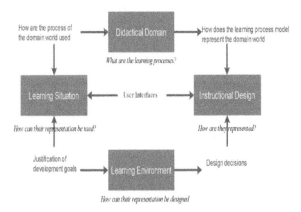

Figure 2.2. *Framework worlds (Bayounes et al. 2012)*

2.3.1. *Didactic domain world*

This is the world of processes taking into account the notion of the process and its nature (Bayounes et al. 2012). The didactic domain world contains the knowledge of the domain about which the proposed process has to provide learning (Bayounes et al. 2012). This world specifies the process nature by reviewing the domain dimensions. The nature view defines three facets, namely pedagogy, learning and process.

The pedagogy facet presents three attributes: the pedagogical orientation, the pedagogical method and the correlation. The different orientations are teacher-directed, learner-directed and teacher–learner negotiated. The pedagogical method specifies the activities of the learning process. It can be classified as direct instruction, indirect instruction, interactive instruction, experiential learning and independent study. The third attribute tests the correlation between the pedagogical and learning processes.

The learning facet presents three attributes: the performance, the learning mode and the learning outcome. The different performance types are remember, use and find (Merrill 1983). These performances are accomplished by three different learning modes, namely accretion, structuring and tuning (Rumelhart and Noorman 1978).

2.3.2. *Instructional design world*

This world deals with the representation of processes by adopting predefined models. It focuses on the description view by respecting the notation conditions and the constraints of representation level. This world includes the notation and the level of process description.

In the notation facet, four attributes are found: the type, the form, the content and the models. The two notation types are standard or proprietary. These types are used by adopting three major forms namely informal, semi-formal and formal. According to the component display theory (Merrill 1983), these forms present four major types of content. These types are fact, concept, procedure and principle. These contents are used to instantiate the domain model, the learner model and the pedagogical model.

Furthermore, the level facet presents three attributes: the granularity, the modularization and the coverage. The different levels of granularity are course, sequence of activities and simple activity. Three types of modularization, namely primitive, generic and aggregation, define these levels (Kumar and Ahuja 2020). In fact, the modularization attribute is used to capture the laws governing the learning process construction by using process modules (Bayounes et al. 2012). This process has been defined differently in different coverage (Bayounes et al. 2014):

The activity: the process is defined as a related set of activities conducted for the specific purpose of product definition.

The product: the process is a series of activities that cause successive product transformations to reach the desired product.

The decision: the process is defined as a set of related decisions conducted for the specific purpose of product definition.

The context: the process is a sequence of contexts causing successive product transformations under the influence of a decision taken in a context.

2.3.3. *Learning environment world*

This world deals with the entities and activities that arise as part of the engineering process itself (Bayounes et al. 2012). It focuses on the design view by describing the implementation of process representation. The design view includes five facets, namely context, construction, optimization, guidance and adaptation.

The first facet defines the type and the form of context. The construction facet deals with four issues: the approaches, the methods, the tools and the techniques of adaptive learning process construction. In a manner analogous to the adaptation spectrum (Patel and Kinshuk 1997), we can organize construction approaches in a spectrum ranging from "low" to "high" flexibility. The construction approach attribute is defined as the approach: ENUM {Rigid, Contingency, On-The-Fly}. These approaches are adopted by using the instantiation, the assembly and the ad hoc method. These methods apply three major techniques (curriculum sequencing, intelligent solution analysis, problem-solving support).

In the optimization facet, four attributes are found: the model, the method, the technique and the parameters of optimization. The different models are map-based, network-based and tree-based. Three types of methods, namely exact, heuristic and meta-heuristic, apply these models. These methods use rule-based and case-based techniques. The achieved educational intentions can be used to implement these optimization techniques.

In addition, the guidance facet defines four attributes: the type, the method, the outcomes and the parameters of guidance. The guidance types are strict and flexible. These types make use of three major methods, namely expository, collaborative and discovery. The outcomes of these methods are activity, resource and intention. To this end, the guidance methods adopt three parameters, namely learning style, cognitive state and teaching style.

Finally, the fourth facet introduces the adaptation by defining five attributes, which are the dimension, the position, the method, the technique and the parameters of adaptation. The three dimensions of adaptation are content, structure and presentation. The position is identified by adopting the adaptation spectrum (Patel and Kinshuk 1997). Each position is satisfied by applying four major methods, namely macro adaptive, aptitude treatment interaction, micro adaptive and constructivist-collaborative (Garcia-barrios et al. 2004). These methods are implemented by four major techniques, which are adaptive interaction, adaptive course delivery, content discovery and assembly and adaptive collaboration support (Paramythis and Loidl-Reisinger 2004). These techniques adopt four major parameters, specifically learning goal, learning history, prior knowledge and system information.

2.3.4. *Learning situation world*

This world supports the scenario view by examining the reason and the rationale of learning process engineering (Bayounes et al. 2012). It describes the organizational environment of the educational process by indicating the purpose and policy process management. The purpose facet includes two attributes: process and learning. In fact, the construction approaches have been designed for different purposes and try to describe the learning process in different attitudes: descriptive, prescriptive and explanatory. The learning purpose defines the level acquired by constructing the learning process. According to Bloom's taxonomy (Forehand 2012), the different levels are knowledge, comprehension, application, analysis, synthesis and evaluation.

The second attribute identifies three policies of process management, particularly evolving, reuse and assessment. This attribute supports the validation of construction process quality. Moreover, since the learning situations change and evolve over time, it is essential that the learning process construction supports these evolutions (Bayounes et al. 2012). As with any process development, the reuse and the assessment, which can occur at any stage of learning process construction and at any level of abstraction and may involve any element of design and/or implementation, are outstanding.

2.4. Comparative study

2.4.1. *Study scope*

Several intelligent tutoring systems have been reported in the literature. Before applying our framework of educational process construction to these systems, Table 2.1 presents a brief description of selected systems. Indeed, this selection is based on:

– various didactic domains;

– different techniques of artificial intelligence;

– several countries from different continents;

– number and quality of related publications.

Intelligent tutoring systems			
ID	**Name**	**Didactic domain**	**Country**
ITS 1	ITS-C	Linguistics	Spain
ITS 2	PEGASE	Virtual reality	France
ITS 3	CIRCISM-Tutor	Medicine	USA
ITS 4	Bits	Programming	Canada
ITS 5	SQL-Tutor	Database	New Zealand

Table 2.1. *Selected ITS*

2.4.2. *Description of systems*

2.4.2.1. *ITS-C (ITS 1)*

The Intelligent Tutoring System based on Competences (ITS-C) extends an ITS by linking the latter and the pedagogical model based on Competency-based Education (Badaracco and Martínez 2013). It adopts Computerized Adaptive Tests (CAT) as common tools for the diagnosis process (Badaracco and Martínez 2013).

2.4.2.2. *PEGASE (ITS 2)*

The PEdagogical Generic and Adaptive SystEm (PEGASE) is used to instruct the learner and to assist the instructor. This system emits a set of knowledge (actions carried out by the learner, knowledge about the field, etc.), which PEGASE uses to make informed decisions (Buche et al. 2010).

2.4.2.3. *CIRCSIM-Tutor (ITS 3)*

CIRCSIM-Tutor is an intelligent tutoring system for teaching the baroreceptor reflex mechanism of blood pressure control to first-year medical students (Glass 2001).

2.4.2.4. *BITS (ITS 4)*

The Bayesian Intelligent Tutoring System (BITS) is a web-based intelligent tutoring system for computer programming using Bayesian technology (Butz et al. 2004).

2.4.2.5. *SQL-Tutor (ITS 5)*

The SQL-Tutor is a problem-solving environment intended to complement classroom instruction, and we assume that students are already familiar with database theory and the fundamentals of SQL (Mitrovic et al. 2000). This ITS adopts constraint-based modeling as a learner modeling approach.

2.4.3. *Specification of approaches*

2.4.3.1. *Approach of ITS 1*

2.4.3.1.1. Nature view

The system adopts the learner-directed orientation by using the indirect and the independent methods (see Table 2.2). The system is used to achieve the intellectual skill and the cognitive strategy by applying the using and the finding performance. In fact, this system supports a tactic and linear process by applying the structuring and tuning modes.

Facet	Definition	
	Attribute	**Values**
Pedagogy	Orientation	{Learner-Directed}
	Method	{Indirect, Independent Study}
	Correlation	{Not Considered}
Learning	Performance	{Use, Find}
	Mode	{Structuring, Tuning}
	Outcome	{Intellectual Skill, Cognitive Strategy}
Process	Type	{Tactic}
	Form	{Linear}

Table 2.2. *Nature view of ITS 1*

2.4.3.1.2. Description view

By adopting the learner model, the pedagogical model and the domain model, the system adopts a standard and formal notation of the learning process (see Table 2.3). This notation is used to describe the required concepts. By using the aggregation, the system uses a description of the activities sequence to cover the desired product.

Facet	Definition	
	Attribute	**Values**
Notation	Type	{Standard}
	Form	{Formal}
	Content	{Concept}
	Models	{Domain Model, Learner Model, Pedagogical Model}
Level	Performance	{Sequence}
	Mode	{Aggregation}
	Outcome	{Product}

Table 2.3. *Description view of ITS 1*

2.4.3.1.3. Design view

By respecting an uncertain and an evolved context, the system supports the contingency approach. This construction approach applies the assembly methods by adopting the problem-solving support technique (see Table 2.4).

The system endorses a network-based model of optimization by defining an exact method. This method is specified by using different rules. Moreover, the system supports a flexible guidance by employing discovery methods to provide the suitable activity. For that, these methods consider the cognitive state. To offer a structure-based adaptation, the system uses the method aptitude–treatment interaction by adopting the technique content discovery and assembly. This technique considers the prior knowledge to satisfy the position system-initiated adaptivity with pre-information to the user about the change.

Facet	Definition	
	Attribute	**Values**
Context	Type	{Uncertain}
	Form	{Evolved}
Construction	Approach	{Contingency}
	Method	{Assembly}
	Technique	{Problem-Solving Support}
Optimization	Model	{Network-Based}
	Method	{Exact}
	Technique	{Rule-Based}
	Parameters	{Not Defined}
Guidance	Type	{Flexible}
	Method	{Discovery}
	Parameters	{Cognitive State}
	Outcome	{Activity}
Adaptation	Dimension	{Structure}
	Position	{System-Initiated Adaptivity with pre-information}
	Method	{Aptitude–Treatment Interaction}
	Technique	{Content Discovery and Assembly}
	Parameters	{Prior Knowledge}

Table 2.4. *Design view of ITS 1*

2.4.3.1.4. Scenario view

To achieve the evaluation level, the system supports a prescriptive process of learning. The reuse and the assessment of the process are considered (see Table 2.5).

Facet	Definition	
	Attribute	**Values**
Purpose	Process	{Prescriptive}
	Learning	{Evaluation}
Policy	Reuse	{True}
	Evolving	{False}
	Assessment	{True}

Table 2.5. *Scenario view of ITS 1*

2.4.3.2. *Approach of ITS 2*

2.4.3.2.1. Nature view

The system adopts the learner-directed and the teacher–learner negotiated pedagogical orientations by implementing the experiential pedagogical methods (see Table 2.6). By considering the pedagogical correlation, the system is used to reach the desired cognitive strategy. This system supports a strategic and linear process of learning through applying the structuring and the tuning modes. These modes are adopted to achieve the using and the finding performance.

Facet	Definition	
	Attribute	Values
Pedagogy	Orientation	{Learner-Directed, Teacher–Learner Negotiated}
	Method	{Experiential}
	Correlation	{Considered}
Learning	Performance	{Use, Find}
	Mode	{Structuring, Tuning}
	Outcome	{Cognitive Strategy}
Process	Type	{Strategic}
	Form	{Linear}

Table 2.6. *Nature view of ITS 2*

2.4.3.2.2. Description view

The system applies a standard and formal notation of the learning process by using the learner model, the domain model and the pedagogical model (see Table 2.7). This notation is used to describe the appropriate procedure. By using the aggregation, the system employs a description of the activities sequence to meditate over the context.

Facet	Definition	
	Attribute	**Values**
Notation	**Type**	{Standard}
	Form	{Formal}
	Content	{Procedure}
	Models	{Domain Model, Learner Model, Pedagogical Model}
Level	**Performance**	{Sequence}
	Mode	{Aggregation}
	Outcome	{Context}

Table 2.7. *Description view of ITS 2*

2.4.3.2.3. Design view

By considering a certain and an evolved context of construction, the system supports the on-the-fly approach. This construction approach applies ad hoc methods by adopting the intelligent solution analysis technique (see Table 2.8).

By adopting the achieved learning intention, the system applies an exact method of optimization, by defining different rules. In addition, the system supports a flexible guidance by using discovery and expository methods to provide the suitable activity. For that, these methods consider the cognitive state. To support a presentation-based adaptation, the system uses the constructivist-collaborative method by applying the technique adaptive collaboration support. This technique considers the learning history to offer the position system-initiated adaptivity with pre-information to the user about the change.

Facet	Definition	
	Attribute	**Values**
Context	**Type**	{Certain}
	Form	{Evolved}
Construction	**Approach**	{On-The-Fly}
	Method	{Ad hoc}
	Technique	{Intelligent Solution Analysis}
Optimization	**Model**	{Not Defined}
	Method	{Exact}
	Technique	{Rule-Based}
	Parameters	{Achieved Learning Intention}
Guidance	**Type**	{Flexible}
	Method	{Discovery, Expository}
	Parameters	{Cognitive State}
	Outcome	{Activity}
Adaptation	**Dimension**	{Presentation}
	Position	{System-Initiated Adaptivity with pre-information to the user about the change}
	Method	{Constructivist-Collaborative}
	Technique	{Adaptive Collaboration Support}
	Parameters	{Learning History}

Table 2.8. *Description view of ITS 2*

2.4.3.2.4. Scenario view

To achieve the evaluation level, the system supports a prescriptive process of learning. The reuse and assessment of the process are considered (see Table 2.9).

Facet	Definition	
	Attribute	Values
Purpose	Process	{Descriptive, Explanatory}
	Learning	{Comprehension, Application}
Policy	Reuse	{False}
	Evolving	{True}
	Assessment	{True}

Table 2.9. *Scenario view of ITS 2*

2.4.3.3. *Approach of ITS 3*

2.4.3.3.1. Nature view

The system adopts the teacher–learner negotiated orientation by using the indirect and interactive pedagogical methods (see Table 2.10). The system is used to reach the desired cognitive strategy and the suitable intellectual skill. For that, the system supports a tactic and linear process of learning by applying the structuring mode. This learning mode is applied to achieve the different types of performance.

Facet	Definition	
	Attribute	Values
Pedagogy	Orientation	{Teacher–Learner Negotiated}
	Method	{Indirect, Interactive}
	Correlation	{Not considered}
Learning	Performance	{Remember, Use, Find}
	Mode	{Structuring}
	Outcome	{Intellectual Skill, Cognitive Strategy}
Process	Type	{Tactic}
	Form	{Linear}

Table 2.10. *Nature view of ITS 3*

2.4.3.3.2. Description view

The system applies a proprietary and informal notation of the learning process by using the learner model and the domain model (see Table 2.11). This notation is adopted to describe the required procedure. By using the aggregation, the system supports an activity description of the learning process.

Facet	Definition	
	Attribute	Values
Notation	Type	{Proprietary}
	Form	{Informal}
	Content	{Procedure}
	Models	{Domain Model, Learner Model}
Level	Performance	{Activity}
	Mode	{Aggregation}
	Outcome	{Activity}

Table 2.11. *Description view of ITS 3*

2.4.3.3.3. Design view

The system adopts the on-the-fly approach by applying ad hoc methods. This approach supports a certain and stable context of construction through the application of the problem-solving support technique (see Table 2.12).

The system supports an exact method by designing different rules of optimization. Moreover, it adopts a flexible guidance by using collaborative methods to provide the suitable activity. For this purpose, these methods consider the cognitive state. In addition, the system provides a content-based adaptation by using the constructivist-collaborative method. By using the technique adaptive collaboration support, the method adopts the prior knowledge to satisfy the position system-initiated adaptivity with pre-information to the user about the change.

Facet	Definition	
	Attribute	**Values**
Context	Type	{Certain}
	Form	{Stable}
Construction	Approach	{On-The-Fly}
	Method	{Ad hoc}
	Technique	{Problem-Solving Support}
Optimization	Model	{Not Defined}
	Method	{Exact}
	Technique	{Rule-Based}
	Parameters	{Not Defined}
Guidance	Type	{Flexible}
	Method	{Collaborative}
	Parameters	{Cognitive State}
	Outcome	{Activity}
Adaptation	Dimension	{Content}
	Position	{System-Initiated Adaptivity with pre-information}
	Method	{Constructivist-Collaborative}
	Technique	{Adaptive Collaboration Support}
	Parameters	{Prior Knowledge}

Table 2.12. *Description view of ITS 3*

2.4.3.3.4. Scenario view

The system supports a prescriptive process of learning to achieve the comprehension level. For that, the assessment and the reuse of the process are considered (see Table 2.13).

Facet	Definition	
	Attribute	**Values**
Purpose	Process	{Prescriptive}
	Learning	{Comprehension}
Policy	Reuse	{True}
	Evolving	{False}
	Assessment	{True}

Table 2.13. *Scenario view of ITS 3*

2.4.3.4. *Approach of ITS 4*

2.4.3.4.1. Nature view

By considering the learner-directed orientation, the system adopts the direct pedagogical methods (see Table 2.14). The system is used to reach the desired cognitive strategy, the required intellectual skill and the appropriate verbal information. For this purpose, the system supports a tactic and linear process of learning. This process applies the different learning modes to achieve the using and the remembering performance.

Facet	Definition	
	Attribute	**Values**
Pedagogy	Orientation	{Learner-Directed}
	Method	{Direct}
	Correlation	{Not considered}
Learning	Performance	{Remember, Use}
	Mode	{Accretion, Structuring}
	Outcome	{Verbal Information, Intellectual Skill, Cognitive Strategy}
Process	Type	{Tactic}
	Form	{Linear}

Table 2.14. *Nature view of ITS 4*

2.4.3.4.2. Description view

The system applies a standard and formal notation of the learning process by using the learner model and the domain model (see Table 2.15). This notation is used to describe the required concepts. The system uses an aggregation or primitive description of learning activity.

Facet	Definition	
	Attribute	**Values**
Notation	Type	{Standard}
	Form	{Formal}
	Content	{Concept}
	Models	{Domain Model, Learner Model}
Level	Performance	{Activity}
	Mode	{Aggregation, Primitive}
	Outcome	{Activity}

Table 2.15. *Description view of ITS 4*

2.4.3.4.3. Design view

The system supports the contingency approach to consider uncertain and stable contexts. This approach applies the instantiation methods of construction by adopting the curriculum sequencing technique (see Table 2.16).

By considering the network-based model, the system adopts an exact method of optimization by defining different rules. Moreover, the system supports a flexible guidance by using expository methods to offer the suitable resource. To this end, these methods consider the cognitive state. To support a structure and content-based adaptation, the system uses the method aptitude–treatment interaction by applying the adaptive course delivery technique. This technique adopts the learning goal and the prior knowledge to offer the position user selection adaptation for system suggested features.

Facet	Definition	
	Attribute	**Values**
Context	Type	{Uncertain}
	Form	{Stable}
Construction	Approach	{Contingency}
	Method	{Instantiation}
	Technique	{Curriculum Sequencing}

Optimization	**Model**	{Network-Based}
	Method	{Exact}
	Technique	{Rule-Based}
	Parameters	{Not Defined}
Guidance	**Type**	{Flexible}
	Method	{Expository}
	Parameters	{Cognitive State}
	Outcome	{Resource}
Adaptation	**Dimension**	{Content, Structure}
	Position	{User Selection Adaptation for system suggested features}
	Method	{Aptitude–Treatment Interaction}
	Technique	{Adaptive Course Delivery}
	Parameters	{Learning Goal, Prior Knowledge}

Table 2.16. *Description view of ITS 4*

2.4.3.4.4. Scenario view

The system supports a descriptive process of learning to achieve the knowledge and the comprehension level. For this purpose, the reuse and the evolving of the process are adopted by the system (see Table 2.17).

Facet	**Definition**	
	Attribute	**Values**
Purpose	**Process**	{Descriptive}
	Learning	{Knowledge, comprehension}
Policy	**Reuse**	{True}
	Evolving	{True}
	Assessment	{False}

Table 2.17. *Scenario view of ITS 4*

2.4.3.5. *Approach of ITS 5*

2.4.3.5.1. Nature view

The system adopts the learner-directed orientation to support the independent study and the indirect pedagogical methods (see Table 2.18). The system is employed to reach the desired cognitive strategy by supporting a tactic and linear process of learning. The latter applies the structuring and tuning modes to achieve the using and the finding performance.

Facet	Definition	
	Attribute	Values
Pedagogy	Orientation	{Learner-Directed}
	Method	{Indirect, Independent Study}
	Correlation	{Not considered}
Learning	Performance	{Use, Find}
	Mode	{Structuring, Tuning}
	Outcome	{Cognitive Strategy}
Process	Type	{Tactic}
	Form	{Linear}

Table 2.18. *Nature view of ITS 5*

2.4.3.5.2. Description view

The system adopts a proprietary and informal notation of the learning process via the implementation of the learner model, the pedagogical model and the domain model (see Table 2.19). This notation is adopted to describe the appropriate procedure.

Facet	Definition	
	Attribute	Values
Notation	Type	{Proprietary}
	Form	{Informal}
	Content	{Procedure}
	Models	{Domain Model, Learner Model, Pedagogical Model}
Level	Performance	{Activity}
	Mode	{Primitive}
	Outcome	{Product}

Table 2.19. *Description view of ITS 5*

2.4.3.5.3. Design view

The system adopts the on-the-fly approach to consider the certain and stable context of construction. In fact, it supports ad hoc methods by applying the intelligent solution analysis technique (see Table 2.20).

The system adopts an exact method by specifying different rules. In addition, the system adopts a flexible guidance by using discovery methods to provide the appropriate resource. To this end, these methods incorporate a cognitive state. Furthermore, the system offers a content-based adaptation by implementing the aptitude–treatment interaction method. This method adopts the adaptive interaction technique in light of the learning history. In fact, this method is used to support the position of system-initiated adaptivity with pre-information to the user about the change.

Facet	Definition	
	Attribute	Values
Context	Type	{Certain}
	Form	{Stable}
Construction	Approach	{On-The-Fly}
	Method	{Ad hoc}
	Technique	{Intelligent Solution Analysis}
Optimization	Model	{Network-Based}
	Method	{Exact}
	Technique	{Rule-Based}
	Parameters	{Not Defined}
Guidance	Type	{Flexible}
	Method	{Discovery}
	Parameters	{Cognitive State}
	Outcome	{Resource}
Adaptation	Dimension	{Content}
	Position	{System-Initiated Adaptativity with pre-information}
	Method	{Aptitude–Treatment Interaction}
	Technique	{Adaptive Interaction}
	Parameters	{Learning History}

Table 2.20. *Description view of ITS 5*

2.4.3.5.4. Scenario view

The system endorses an explanatory process of learning to achieve the comprehension and the application level. For this purpose, the assessment and the evolving of the process are supported by the system (see Table 2.21).

Facet	Definition	
	Attribute	**Values**
Purpose	Process	{Explanatory}
	Learning	{Comprehension, Application}
Policy	Reuse	{False}
	Evolving	{True}
	Assessment	{True}

Table 2.21. *Scenario view of ITS 5*

2.4.4. *Study results and discussion*

The framework analysis identifies the following main drawbacks of existing adaptive construction approaches. It should be noted that construction approaches are not sufficiently automated to automatically derive the achievement of objectives from the strategic process.

For the nature view, the different systems do not adopt all pedagogical methods to achieve the different types of process outcomes. In fact, the outcomes attitude and motor skill are not taken into consideration. In addition, the different systems support the tactic and linear processes of learning.

For the description view, most of the systems try to support the learner model, the domain model and the pedagogical model by applying proprietary notation, which can manifest some standardization problems. This notation is not used to specify a process description by respecting the activity level. Indeed, the context coverage is not supported by the learning process description.

For the design view, the uncertain and the evolved context of construction are not supported by the systems. The different on-the-fly approaches do not support a combined construction technique of learning processes to implement the instantiation methods. Most of these approaches are not implemented by applying a MAP-based model of optimization. For that, the educational intentions are not defined as optimization parameters. The construction approaches themselves are not sufficiently guided by considering the different methods and parameters. In fact, the flexible guidance is most used by the expository methods. However, this flexible guidance does not adopt the intention as outcome by considering the learning style and the teaching style.

Finally, the different systems do not apply micro methods by respecting system information to support process adaptation.

Therefore, the comparative study identifies the following limits:

– the adaptive ability of most of these ITS is limited to the content. In fact, the process-oriented adaptation is not supported by the different systems;

– the consideration of the educational process as a strategic and not linear one is limited;

– the flexible guidance of educational processes does not adopt the intention as outcome by considering the learning style and the teaching style;

– the different systems did not respect the evolving process policy.

2.5. Conclusion and future works

This work presents a new multi-level view of educational process definition. This view is introduced to depict a comparison framework, which has allowed identification of the characteristics and drawbacks of some existing construction approaches supported by ITS. This framework considers adaptive educational process engineering from four different but complementary points of view (Nature, Description, Design and Scenario). Each view allows us to capture a different aspect of learning process engineering (Bayounes et al. 2019).

As a matter of fact, the framework is applied to respond to the following purposes: to have an overview of the existing construction approaches, to identify their drawbacks and to analyze the possibility of proposing a better approach.

In order to study, understand and classify different construction approaches of educational processes in their diversity, we selected different ITS from five continents. After the use of the proposed framework, the first view considers that the different systems do not adopt all pedagogical methods. The second view identifies that the context coverage is not depicted. In fact, the uncertain context of construction is not supported by most of the systems. In addition, the evolving process policy is not respected.

The analysis of existing approaches definition identifies the lack of an adaptive and guided construction of educational processes that embody the new technologies of artificial intelligence. It should be noted that the majority of ITS do not respect the correlation between pedagogical and learning processes. Our future work will include the learner's motivation (Dunn and Kennedy 2019) as a success factor of educational process construction.

2.6. References

Badaracco, M. and Martínez, L. (2013). A fuzzy linguistic algorithm for adaptive test in intelligent tutoring system based on competences. *Expert Systems with Applications*, 40(8), 3073–3086.

Bayounes, W., Saâdi, I.B., Kinshuk, K., Ben Ghézala, H. (2012). Towards a framework definition for learning process engineering supported by an adaptive learning system. *Proc. ICTEE Conference*, Amritapuri, India, 366–373.

Bayounes, W., Saâdi, I.B., Kinshuk, K., Ben Ghézala, H. (2013). An intentional model for learning process guidance in adaptive learning system. *Proc. 22nd IBIMA Conference*, Rome, Italy, 1476–1490.

Bayounes, W., Saâdi, I.B., Kinshuk, K., Ben Ghézala, H. (2014). An intentional model for pedagogical process guidance in adaptive learning system. *Proc. 23rd IBIMA Conference*, Valencia, Spain, 1211–1227.

Bayounes, W., Saâdi, I.B., Ben Ghézala, H. (2019). Educational processes' guidance based on evolving context prediction in intelligent tutoring systems. *Universal Access in the Information Society*, 8(68), 1–24.

Bayounes, W., Saâdi, I.B., Ben Ghézala, H. (2020). Definition framework of educational process construction supported by an intelligent tutoring system. *Proceedings Multi-Conference OCTA (Organization of Knowledge and Advanced Technologies)*, Tunis, Tunisia [Online]. Available at: https://multiconference-octa.loria.fr/multiconference-program/.

Buche, C., Bossard, C., Querrec, R., Chevaillier, P. (2010). {PEGASE}: A generic and adaptable intelligent system for virtual reality learning environments. *IJVR*, 9(2), 73–85.

Butz, C., Hua, S., Maguire, R.B. (2004). Bits: A Bayesian intelligent tutoring system for computer programming. *WCCCE '04*, 179–186.

Chang, M., D'Aniello, G., Gaeta, M., Orciuoli, F., Sampson, D., Simonelli, C. (2020). Building ontology-driven tutoring models for intelligent tutoring systems using data mining. *IEEE Access*, 8, 48151–481621.

Dunn, T.J. and Kennedy, M. (2019). Technology enhanced learning in higher education; Motivations, engagement and academic achievement. *Computers & Education*, 137, 104–113.

Forehand, M. (2012). Bloom's taxonomy. In *Emerging Perspectives on Learning, Teaching, and Technology*, Orey, M. (ed.). CreateSpace, Scotts Valley, CA, USA.

Garcia-Barrios, V.M., Mödritscher, F., Christian, G. (2004). The past, the present and the future of adaptive e-learning: An approach within the scope of the research project AdeLE. *Proc. ICL Conference*, Villach, Austria, 1–9.

Glass, M. (2001). Processing language input in the CIRCSIM-tutor intelligent tutoring system. *Artificial Intelligence in Education*, 210–221.

Kumar, A. and Ahuja, N.J. (2020). An adaptive framework of learner model using learner characteristics for intelligent tutoring systems. In *Intelligent Communication, Control and Devices*, Choudhury, S., Mishra, R., Mishra, R.G., Kumar, A. (eds). Springer, Singapore.

Merrill, M.D. (1983). Component display theory. *Instructional Design Theories and Models: An Overview of their Current Status*, Reigeluth, C.M. (ed.). Routledge, New York, USA.

Mitrovic, A., Martin, B., Mayo, M. (2000). Using evaluation to shape ITS design: Results and Experiences with SQL-Tutor. *Int. J. User Modeling and User-Adapted Interaction*, 12(2–3), 243–279.

Olesea, C. (2019). The concept of personal learning pathway for intelligent tutoring system GeoMe. *Computer Science Journal of Moldova*, 27(3), 355–367.

Paramythis, A. and Loidl-Reisinger, S (2004). Adaptive learning environments and e-learning standards. *Electronic Journal of Elearning*, 2(1), 181–194.

Patel, A. and Kinshuk, K. (1997). Intelligent tutoring tools in a computer integrated learning environment for introductory numeric disciplines. *Innovations in Education & Training International*, 34(3), 200–207.

Rathore, A.S. and Arjaria, S.K. (2020). Intelligent tutoring system. In *Utilizing Educational Data Mining Techniques for Improved Learning: Emerging Research and Opportunities*, Bhatt, C., Sajja, P., Liyanage, S. (eds). IGI Global, Hershey, PA, USA.

Rolland, C. (1998). A comprehensive view of process engineering. *Proc. 10th CAISE Conference*, Pisa, Italy, 1–24.

Rumelhart, D.E. and Noorman, D.A. (1978). Accretion, tuning, and restructuring: Three modes of learning. In *Semantic Factors in Cognition*, Cotton, W. and Klatzky, R. (eds). Erlbaum, Hillsdale, NJ, USA.

Tzanova, S. (2020). Revolution by evolution: How intelligent tutoring systems are changing education. In *Revolutionizing Education in the Age of AI and Machine Learning*, Habib, M. (ed.). IGI Global, Hershey, PA, USA.

Multi-Criteria Decision-Making Recommender System Based on Users' Reviews

3.1. Introduction

Due to the frequent accumulation of web resources, finding useful and relevant information is becoming an increasingly difficult task to manage. RSs have been integrated into the online information systems (e.g. e-commerce platforms, social media sites, etc.) to solve the increasing information overload problem by providing users with new items that may interest them. Most of the existing recommendation systems (Al-Ghuribi and Mohd Noah 2019) focus on the historical data of users' preferences on the overall representation of items considering that all its criteria have the same importance. This restricted vision cannot present the evaluation of the users' interests correctly. For instance, when choosing a restaurant, we can consider food quality as a primary criterion and the remaining ones (e.g. service, ambiance and price) may not be a big concern for them even if they are highly rated. Recently, many research works (Yang et al. 2016; Zheng 2019; Briki et al. 2020) have proposed applying multi-criteria decision-making (MCDM) methods to consider the multi-faceted representation of items and users' preference in the recommendation generation process. They prove a considerable progress in the recommendation performance. Most of these works consider the numerical ratings on the different criteria of items to generate recommendations. However, the available databases correspond,

Chapter written by Mariem BRIKI, Sabrine BEN ABDRABBAH and Nahla BEN AMOR.

generally, to a huge volume of unstructured data in the form of textual data derived from several sources (e.g. comments from shared publications on social networks, replies to registration questionnaires, etc.). The users' reviews, comments and experiences may be a powerful source of information that can be exploited in RS in order to model the users' profiles. At this point, text mining analysis techniques (e.g. categorization, entity extraction, sentiment analysis and natural language processing) are used to find the hidden knowledge in text content and revealing useful patterns, trends and insights.

Our objective in this chapter is to improve the recommendation task by using text mining techniques in order to capture the multi-criteria of users' interests from the users' reviews. The main idea consists of defining the criteria of the items, and then creating the corpus of information of each criterion. On this basis, we analyze the active user's reviews in order to identify their primary criterion. Finally, we analyze all the reviews of each unseen item in order to capture the ones that match the active user's primary criterion.

Throughout this work, we have implemented two algorithms, namely the primary criterion-based recommendation system (PCRS) and the multi-criteria text mining-based recommendation system (MCTMRS). The PCRS consists of taking advantage of text mining techniques to analyze the active user's reviews and to determine their primary criterion. The MCTMRS consists of using the weighted sum method (WSM) (Pohekar and Ramachandran 2004) as it is the most commonly used method in the multi-criteria decision-making (MCDM) field to compute the best decision alternative with respect to the weight of each criterion.

The rest of this chapter is organized as follows: section 3.2 is devoted to the multi-criteria decision-making methods, section 3.3 describes the basics of RS and related works, section 3.4 details the proposed solution and section 3.5 presents the conclusions emerged from the experiments.

3.2. Multi-criteria decision-making

Multi-criteria decision-making (MCDM) is a branch of operational research dealing with finding optimal results in complex scenarios including various indicators, conflicting objectives and criteria. Applications of

MCDM include areas such as integrated manufacturing systems, evaluations of technology investment, and water and agriculture management. Several methods have been introduced in the multi-criteria decision-making field to enable users to make decisions with respect to different features/parameters. The most commonly used methods are the following:

– The weighted sum method (WSM) (Pohekar and Ramachandran 2004) is used for evaluating a number of alternatives in accordance with the different criteria, which are expressed in the same unit. For instance, let M be the set of alternatives and N be the set of decision criteria; the WSM score is computed as the weighted sum of the performance value of each criteria j on the alternative i. The best alternative is the one having the highest WSM score.

$$A^*_{WSM} = Max \sum_i^j \ a_{ij}w_j \qquad\qquad [3.1]$$

where i = 1, 2 ... M, j = 1, 2 ... N, a_{ij} is the performance value of the ith alternative in terms of the jth criterion and w_j is the relative weight of importance of the jth criterion.

– The weighted product method (WPM) (Pohekar and Ramachandran 2004) selects the most important alternative from a set of alternatives M with respect to a set of criteria N. Every alternative A_K is compared with the other alternative A_L by multiplying the powered ratios of the first and the second performance values with respect to every criterion. If $R(\frac{A_k}{A_L})$ is greater than 1, then the alternative A_K is more desirable than the alternative A_L. A_K is the best alternative if it is greater than all the other alternatives. Formally, this product is computed as follows:

$$R(A_k/A_L) = \prod_{j=1}^{N} \ \left(\frac{a_{Kj}}{a_{Lj}}\right)^{w_j} \qquad\qquad [3.2]$$

where a_{Kj} is the performance value of the kth alternative in terms of the jth criterion and w_j is the relative weight of importance of the jth criterion.

– The analytical hierarchy process (AHP) (Saaty 1985) constructs at first a logical hierarchy of criteria by defining the objective at the top of the hierarchy, criteria and sub-criteria at levels and sub-levels of the hierarchy, and decision alternatives at the bottom of the hierarchy. Each alternative has its own values of associated criteria. Then, a pairwise comparison matrix is created to give the relative importance of various criteria with respect to the main objective. In the final step of the process, numerical priorities are calculated for each of the decision alternatives.

3.3. Basics of recommendation systems and related work

We recall in this section the basic concepts of recommender systems. Then, we expose text mining-based recommendations and their related works.

3.3.1. *Recommender systems*

Due to the huge amount of resources accessible via the Internet, finding relevant information becomes a challenging task. As a response, recommender systems (RSs) have been introduced to achieve personalization and increase users' satisfaction by recommending items that fit their needs and tastes. These systems have been widely used in various domains and diverse applications and have drawn increasing attention from different research communities such as machine learning, electronic commerce and information retrieval. The main task of recommender systems is to analyze as much as possible users' feedback in order to learn the users' preferences. RSs rely on different types of input data such as:

– *explicit feedback*, which is the information we obtain by directly questioning the user about the proposed items;

– *implicit feedback*, which is the information we obtain by analyzing the behavior of the user such as clicks/queries/watches.

The output of RS can be twofold (Vozalis and Margaritis 2003):

– a *prediction* expressed as a numerical value, $r_{u,j}$, which represents the anticipated opinion of active user u for item j. This predicted value should necessarily be within the same numerical scale (e.g. 1 – bad and 5 – excellent) as the input referring to the opinions provided initially by active user u. This form of recommendation system's output is also known as individual scoring;

– a *recommendation* expressed as a list of N items, where $N \leq n$, which the active user is expected to like the most. The usual approach in that case requires this list to include only items that the active user has not already purchased, seen or rated. This form of recommendation system's output is known as top-N recommendation or ranked scoring.

The use of efficient and accurate recommendation methods is very important for a system that will provide good and useful recommendations to its users. This process explains the importance of understanding the features and the potentials of the different recommendation approaches. The recommendation methods are classified into three main classes, including *content-based*, *collaborative filtering* and *hybrid*. First, the *collaborative filtering approach* (Breese et al. 1998) relies on a matrix of user-item ratings to predict unknown matrix entries, and thus to decide which items to recommend. There are two main CF approaches: *item-based* (Sarwar et al. 2001), where the recommendations are generated based on similar items that the active user has liked in the past, and *user-based* (Shardanand, and Maes 1995), where the recommendations are computed based on the opinions of similar users. Second, the *content-based approach* (Soboroff and Nicholas 1999) uses the contents of the items that the active user has liked in the past and suggests items having similar features to them. And third, the *hybrid approach* (Chu and Tsai 2017) combines *content-based filtering* and *collaborative filtering* in order to take advantage of the representation of the content and the rating information of users to produce recommendations.

Most of the RS algorithms consider the quantitative preferences of users, which are generally expressed on a numerical scale. However, the quantitative notes may not be consistent. In fact, they can be affected by many key factors including the user's mood when one user may react differently with the same item according to their situation, the limited scale when the user may give the same rating to two items that they appreciate differently since the scale of possible values is generally reduced, etc. Thus, numerical data of users are not reliable and cannot represent the precise degree of users' liking. For these reasons, many works highlight the use of textual data as it may be a strong support to consider in the recommendation process.

3.3.2. *Text mining-based recommendation systems*

Text mining is a process to extract interesting and significant patterns to explore knowledge from textual data sources. It is a multidisciplinary field based on information retrieval, data mining, machine learning, statistics, and computational linguistics (Talib et al. 2016). Several works have used the text mining techniques in RS to analyze the users' personal reviews and online behavior and capture their interests. For instance, Li et al. (2015) proposed an RS based on opinion mining. The idea consists of taking advantage of the text mining methods in order to detect the opinion-related information from the massive users' reviews and find the most suitable products for customers. Ziani et al. (2017) presented a basic tool that can be used to analyze Algerian reviews and comments and detect their polarity in order to generate meaningful recommendations for users.

Ganu et al. (2009) proposed a restaurant recommendation system that exploits the users' comments in order to find out their relative topics and sentiment information. In fact, they used the regression model to estimate scaled sentiment points instead of the two bipolar classes: positive or negative. The authors in Lin et al. (2015) developed a personalized hotel recommendation approach based on both textual and contextual data. They identified the user's preferred aspects through tracking the browsing behavior on the mobile devices. The authors in Naw and Hlaing (2013) proposed a content-based recommendation method that helps users who want to buy a car by providing them with relevant car information. They implemented a key extraction algorithm to identify the items' features from the users' reviews.

All these works have exploited users' reviews and the textual-related information in order to understand its semantics, and then capture the users' profiles. However, they are unable to deeply treat the multi-criteria aspect of users' preferences that is a crucial feature to consider when building user profiles and generating recommendations.

3.3.3. *Multi-criteria recommender systems*

The multi-criteria aspect is introduced in the context of RS to detect the multi-faceted representation of users' interests and to correctly build the users' profiles. In fact, the MCDM methods have been used to make more effective recommendations, in numerous application domains. There are two categories of multi-criteria recommender systems. The first category focuses on using the multi-criteria ratings in order to model a user's utility for an item as a vector of ratings along several criteria.

For instance, Hdioud et al. (2017) proposed an objective weight determination method called CCSD to determine the criteria with the most important impact on the decision-making process. The main idea consists of computing the overall assessment value of each item as the weighted sum of the performance value over all the criteria; then, they removed one attribute at a time from the set of criteria and considering its correlation with the overall assessment of items without the inclusion of this removed attribute. The alternative with a small correlation should be given a very important weight. In Zheng (2019), the authors developed a utility-based multi-criteria recommender system that computes a utility function of each item based on the multi-criteria ratings. The utility score is defined using the similarity between the vector of the user's evaluations and the vector of the user's expectations. In fact, the user's expectation vector is learned using three optimization learning-to-rank methods (i.e. pointwise ranking, pairwise ranking and listwise ranking).

The second category focuses on analyzing the unstructured textual reviews of users and items' descriptions to determine the criteria of users' preferences and to generate recommendations.

In Yashvardhan et al. (2015), the authors implemented a multi-criteria recommendation system for hotel recommendations. The proposed algorithm used various NLP approaches on a hotel review corpus and the user's reviews in order to construct a user-item-feature database. The latter is exploited to compute the rating of a hotel from previous users' preferences with respect to different features. In Yang et al. (2016), the authors proposed a collaborative filtering framework that incorporates both user opinions and preferences on different aspects. The first step consists of analyzing the users' reviews to extract the opinion words and then computing a score for each aspect. In the second step, the aspect weight is computed using a tensor factorization approach. Finally, the scores and the weights of multiple aspects are used to infer the overall rating, which measures the user's satisfaction about the item. In Ebadi and Krzyzak (2016), the authors proposed a highly accurate hotel recommender system, which uses the multi-aspect rating system and the large-scale data of different types to suggest hotels for users. In fact, it employs the NLP and the topic modeling techniques to assess the sentiment of the users' reviews and extract implicit items criteria. Then, each aspect of the subject problem is treated using a specific sub-recommender system.

Most of these works have shown that when exploiting the textual data to determine the users' profiles, the recommendation accuracy can be improved. However, most of them focus on the reviews given, one for each specific feature/criteria to infer the criteria weights; then, they use the users' preferences to compute the overall rating, which cannot accurately reflect the users' preferences. In fact, the users' reviews could not contain positive terms/words because they are not completely satisfied by the most interesting criteria for them. Hence, we need to dive deep into the user's reviews to determine the most interesting criteria for them and to exploit the users' reviews to capture the item that meets with their expectation.

3.4. New multi-criteria text-based recommendation system

Our objective is to use the text mining techniques on the users' reviews to compute the users' interests on different items' criteria and to find out the items that meet the selected criteria. We assume that by analyzing the users' reviews, we could reveal much hidden information about users' preferences on multiple criteria/features and detect the prinicpal/most interesting criterion for them. To this end, we propose two algorithms, the primary criterion-based recommendation system (PCRS) and the multi-criteria text mining-based recommendation system (MCTMRS). Both these algorithms should proceed with a pre-processing phase that consists of defining the decision criteria (e.g. the decision criteria for a mobile phone are the price, the storage space, the camera quality and the look/design) and creating the information corpus of each criterion. The information corpus is a collection of linguistic data, words and sentences used in the context of the relative criterion. It is commonly used to perform a statistical analysis for different concepts. For example, for the criteria "service", we created a list of words (i.e. verbs, adjectives, etc.) that can be related or that can describe the relation between clients and persons who work in a restaurant.

3.4.1. *Primary criterion-based recommendation system*

This algorithm is based on two main steps. The first step consists of using the information corpus of each criterion and the active user's reviews to compute the user's interests on each criterion. In this algorithm, we will consider just the most interesting criterion for the user, the so-called primary criterion. We assume that by managing only the primary criterion to create the user's profile, a more focused representation of the user's preferences is provided.

Formally, given the users' reviews, we start by putting all the active user's comments in one sentence denoted as AUR. The AUR is defined as the union of all the active user's reviews. Formally, AUR is:

$$AUR = \cup_{j \in t} \quad AUR_j \qquad [3.3]$$

where t is the number of the active user's reviews and j is the current review.

Then, to find the primary criterion of the active user, we compute the percentage of the existence of each criterion i in the active user's reviews denoted by $UPEC_i$. The criterion with the highest $UPEC_i$ value is selected as the primary criterion of the active user. Formally, $UPEC_i$ is:

$$UPEC_i = \frac{AUR \cap CC_i}{\sum_{b \in n} \quad AUR \cap CC_b} \qquad [3.4]$$

where $b = 1, \ldots n$, n is the number of criteria, i is the current criterion and CC_i is the information corpus of the current criterion.

The second step consists of analyzing the users' reviews of all the unseen items to capture the ones that may satisfy the primary criterion of the active user. Formally, let m be the number of all the items not yet seen by the active user. Given the users' reviews, we put all the comments of each unseen item in one sentence denoted as IR_k, which is defined as the union of all the reviews given to item k. Formally, IR_k is:

$$IR_k = \cup_{j \in s} \quad R_{k,j} \qquad [3.5]$$

where s is the number of all the reviews to item k and $R_{k,j}$ is the jth review to item k.

Finally, to find out the items matching the primary criterion of the active user, we compute the percentage of each criterion i in the review content of the unseen item k denoted by $IPECI_{k,i}$. Formally, $IPECI_{k,i}$ is:

$$IPECI_{k,i} = \frac{IR_k \cap CC_i}{\sum_{b \in n} \quad IR_k \cap CC_b} \qquad [3.6]$$

where $b = 1, ... n$, n is the number of criteria, i is the current criterion and k is the current unseen item.

The items having the highest IPEC value with the active user's primary criterion are then selected for the recommendation list. To better explain this, we have summarized the process of this solution in Figure 3.1.

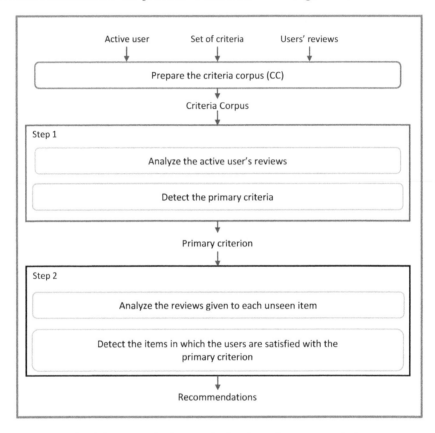

Figure 3.1. *Main steps of primary criterion-based recommendation system. For a color version of this figure, see www.iste.co.uk/sidhom/systems.zip*

Algorithm 3.1 presents the pseudocode of the *primary criterion-based recommendation system* algorithm.

Algorithm 3.1. Primary criterion-based recommendation procedure

---

Require: Users' reviews, information corpus of each criterion
Ensure: Recommendation of items R
  1. $UPEC \leftarrow$ getCriteriaPourcentage($AUR, CC_{i=1..n}$)
  2. **for** (each item $k \in$ m) **do**
  3. $IPEC \leftarrow$ getCriteriaPourcentage($IR_k, CC_{i=1..n}$)
  4. TopU, TopI $\leftarrow 0$
  5. maxU $\leftarrow$ max($UPEC$)
  6. maxI $\leftarrow$ max($IPEC$)
  7. **if** (Index(maxU) is in Index(maxI) **then**
  8. R $\leftarrow$ R + i
  9. **end if**
  10. **end for**
  11. Sort(R)

	Review
Review 1	Very nice discovery. Pleasant cuisine and a warm welcome. Originality of the champagne glasses and delicacy of the minced chicken with asparagus. All accompanied by fries.
Review 2	The welcome is always warm and the dishes well chosen. The menu is a bit expensive compared to the usual restaurant prices. We will come back
Review 3	You have to find the powered farm in Herlies in the north. A real farm of quality products and traditional menus. The prices are reasonable. The welcome is warm and friendly. We spend a good time in an airy and well-decorated space.

Table 3.1. *Reviews of Anthony Rodreguez of Example 3.1*

EXAMPLE 3.1.– Let us consider the following example of the users' reviews given to the restaurants to illustrate the different steps of the proposed PCRS. Obviously, a restaurant can be evaluated with respect to multiple criteria such as service, ambiance, food quality, view and price. We start by preparing the corpus of information of each criterion. For example:

CC_{food}={bowl of soup, mixed green salad, grilled, fried, chicken, salmon, beef, seafood, fruit, ice cream, apple pie, etc.}

CC_{price}={cost, expensive, tips, bills, charge, pay, cash, credit card, etc.}.

The next step consists of analyzing the AUR, which is defined as the union of all the active user's reviews. Table 3.1 shows all the reviews of Anthony Rodreguez. The UPEC of criterion service (respectively service, price, view and ambiance) is 22% (respectively 8%, 6%, 3% and 3%). Thus, the primary criterion for Anthony is the **food quality**. In the last step, we compute the IPEC of each unseen item and we recommend to Anthony "212 steak house" and "kobe beef" restaurants, as the users were very satisfied with their food quality.

3.4.2. *Multi-criteria text mining-based recommendation system*

The multi-criteria text mining-based recommendation system uses the text mining techniques on both the active user's reviews to detect the weight of importance of each criterion and the users' reviews of each unseen item to capture the score of satisfaction of users for each criterion. Then, a multi-criteria decision-making method is applied to compute the overall rating for all the unseen items with respect to multi-criteria preferences. The items with the highest ratings are returned to the user. In this work, we use the *weighted sum method* (WSM) as it is the most commonly used multi-criteria decision-making method.

Algorithm 3.2 presents the pseudo-code of the multi-criteria text mining-based recommendation system algorithm.

Algorithm 3.2. Multi-criteria text mining-based recommendation procedure

---

Require: Users' reviews, information corpus of each criterion
Ensure: Recommendation of items R
    1. $UPEC \leftarrow$ getCriteriaPourcentage($AUR,CC_{i=1..n}$)
    2.   Weight $\leftarrow$ transform_into_weights($UPEC$)
    3.   **for** (each item $k \in$ m) **do**
    4. $IPEC \leftarrow$ getCriteriaPourcentage($IR_k,CC_{i=1..n}$)
    5.   Score $\leftarrow$ transform_into_scores($IPEC$)
    6. $A_{WSM} \leftarrow \sum_{i=1..n}$ ($Score_i * Weight_i$)
    7.   **end for**
    8.   R $\leftarrow$ Max(A_{WSM})

3.5. Experimental study

To evaluate the accuracy of the proposed algorithm, we conduct experiments on a real-world dataset, namely TripAdvisor. All the experiments are implemented in Python 3.7 on a Windows 7-based PC with an Intel Core i3 processor having a speed of 2.40 GHz and 4 GB of RAM and compiled in Eclipse framework. In the following, we first describe the dataset and the evaluation metrics. Then, we present the experimental results.

3.5.1. *Dataset and metrics*

TripAdvisor is an American website that offers tourists advice from consumers on hotels, restaurants, cities and regions, places of leisure, etc. It also provides accommodation and airline ticket booking tools that compare hundreds of websites to find the best deals. Each week hundreds of users visit TripAdvisor to rate and receive recommendations for new restaurants. The users' preferences can be expressed in two forms: the explicit ratings and the unstructured textual reviews.

We start by using the WebCrawler technique to extract the data used to test the performance of our proposed algorithms. In fact, this technique is generally designed to explore the Web. This robot is generally designed to collect resources (e.g. web pages, images, videos, Word documents, PDF, etc.) and allows a search engine to index them.

The extracted data is saved in an Excel file. In the following, we present the textual and the statistical description of the data:

1) **Textual description of the database**: the database extracted is a corpus of comments on restaurants in the Lille region. This dataset contains 47,168 comments and ratings collected from 10,000 users of the site on 1,000 restaurants, from 18/11/2007 to 01/11/2016. Five criteria are considered, including food, service, ambience and price. The rating is given in five discrete levels from 1 to 5.

2) **Statistical description of the database**: a statistical description allows us to better analyze the database by providing statistics that can enable us to better align our contribution. Table 3.2 presents a statistical description of the TripAdvisor dataset.

Number of distinct users	10,000
Number of distinct restaurants	1,000
Average of reviews' number per user	5
Average of reviews' number per restaurant	5
Number of users with 1 review	4,241
Number of users with 3 reviews	1,410
Number of users with 4 reviews	2,115
Number of users with more than 4 reviews	39,362

Table 3.2. *Description of TripAdvisor dataset*

The dataset has been divided into two subsets: training set and testing set. In the training set, we are interested in the users' reviews to detect the primary criterion of the active user, detect the items that match this criterion and finally generate recommendations. The testing set is used to evaluate the effectiveness of the generated recommendations with regard to the real defined rating of the user. To this end, we first split the data chronologically into testing (20% of the recent instances) and training set (the remaining old instances).

3.5.2. *Evaluation metrics*

Recommendation systems research has used several types of measures for evaluating the quality of the provided recommendation. The evaluation metrics can be categorized into three classes, including predictive accuracy metrics, classification accuracy metrics and rank accuracy metrics (Herlocker et al. 2004). In this work, we are interested in classifying metrics because they are most commonly used in natural language processing and easier to interpret directly (Herlocker et al. 2004). In fact, these metrics measure how many times a recommendation system makes correct or incorrect decisions about whether an item is good. Table 3.3 shows the possible categorization of items, where N is the number of items in the database and how the items are categorized.

	Suggested	Non-suggested	Total
Relevant	$N_{r,s}$	$N_{r,n}$	N_r
Irrelevant	$N_{i,s}$	$N_{i,n}$	N_i
Total	N_s	N_n	N

Table 3.3. *Items' categorization*

We can conclude that recommended items can be either successful recommendations (i.e. relevant) or unsuccessful recommendations (i.e. non-relevant) and the relevant items can be either suggested in the recommendation list or not. Precision, recall and F-measure (Basu et al. 1998) are the most popular metrics for evaluating information retrieval systems.

– *Precision* is used to evaluate the validity of a given recommendation list, and it is defined as the ratio of relevant items selected by the active user relative to the number of items recommended to them:

$$P = \frac{N_{r,s}}{N_s}$$ [3.7]

– *Recall* computes the portion of favored items that were suggested for the active user relative to the total number of the objects actually collected by them:

$$R = \frac{N_{r,s}}{N_r}$$ [3.8]

– *F-measure* is a measure of a statistic test accuracy. It considers both precision and recall measures of the test to compute the score. We could interpret it as a weighted average of the precision and recall, where the best F-measure has its value at 1 and worst score at the value 0:

$$F = 2 * \frac{P*R}{P+R}$$ [3.9]

3.5.3. *Experimental protocol*

We used the text mining techniques to clean and transform the users' reviews as follows:

1) eliminate special characters from user comments;

2) create a list of stop words and add an extension. We use the latter to eliminate the stop words from the comments of the users;

3) concatenate all the comments of a single user/item in one sentence;

4) use the tokenization library to have a list of words that builds a sentence;

5) use the stemmatization library to extract the origin of the words from the obtained sentence.

Once the comments are extracted and cleaned, we move to the exploitation step, which consists of determining the user's interests on the different criteria by computing how many words and terms contain the reviews from each criterion corpus.

We choose to compare our proposed algorithms with the multi-criteria algorithms, including the multi-criteria weighted average-based recommendation system (Lee and Teng 2007) that uses users' ratings in the multi-criteria aspect to provide a recommendation and reviews the similarity-based recommendation system (Naw and Hlaing 2013) that uses text mining techniques and the Jaccard similarity measure in order to find similarity between reviews to make the recommendation. To go further, we perform an experiment to test the effect of the sparse data on the effectiveness of all the above algorithms. We tested with different numbers of comments that each user has made (i.e. 3, 4 and more than 4 comments). The results of the different algorithms are listed in Tables 3.4 and 3.5.

3.5.4. *Experimental results*

It can be noted from Table 3.4 that the recommendations performance of the criterion's weighted average-based recommendation system is the worst. In fact, the users should give ratings for each feature of the item regardless of whether they are interested in the features or otherwise. Moreover, the numeric scale cannot be very reliable since it can be interpreted differently. This confirms what is stated in Lee and Teng (2007). Researchers note that the algorithm that is based on computing the average of the criteria is defined as a traditional way to rank multiple items.

Comparing the reviews similarity-based recommendation system with the criterion's weighted average-based recommendation system, we record an improvement that can be explained by the potential of the textual data (e.g. opinions and reviews) in the users' preferences determination.

Table 3.4 shows that this algorithm has the best results regarding the F-measure evaluation metric than all the other algorithms. This improvement is due to the exploitation of the unstructured textual data not only to detect the active user's interests on different criteria but also to detect the items that satisfy the multi-criteria preference of the active user. This can only be a good sign towards the use of text mining techniques and multi-criteria decision-making aspect. Table 3.5 shows that the greater the number of reviews to be analyzed for a user, the better the quality of the recommendation. We could mention that the algorithm multi-criteria text mining-based recommendation system is always the most efficient even with sparse data.

Algorithm	Precision	Recall	F-measure	Response time (sec)
Weighted average	0.034	0.035	0.021	0.265564
Sentence similarity	0.048	0.113	0.040	0.291961
PCRS	0.223	0.753	0.218	158.294949
MCTMRS	0.232	0.772	0.235	183.132370

Table 3.4. *Evaluation metrics results for 1,000 users*

Algorithm	3 reviews	4 reviews	>= 5 reviews
Weighted average	0.026	0.054	0.133
Sentence similarity	0.065	0.098	0.171
PCRS	0.211	0.226	0.270
MCTMRS	0.223	0.221	0.290

Table 3.5. *F-measure results depending on the number of users' reviews*

3.6. Conclusion

In this chapter, we investigate the effects of both the multi-criteria aspect of users' interests and the users' review content analysis on understanding the users' behaviors. The proposed solution (Briki et al. 2020) is implemented based on three steps: the first step consists of the use of information extraction techniques to obtain users' comments from the data source and to create the information corpus of each criterion. The second step consists of the use of text mining techniques to clean and to analyze the users' reviews. The third step consists of identifying the primary criterion of the active user and generate recommendations accordingly. In this work, we have been through two improvement versions that we tested on a real database extracted from the TripAdvisor website. The experimental study shows that our proposed solution provides motivating results compared with the existing works. Future research consists of exploiting the ontology to describe the multi-criteria content and their relationships. The latter can be used to analyze in depth the users' reviews and to increase the performance of recommendation systems.

3.7. References

Al-Ghuribi, S.M. and Mohd Noah, S.A. (2019). Multi-criteria review-based recommender system-the state of the art. *IEEE Access*, 7, 169446–169468.

Basu, C., Hirsh, H., Cohen, W. (1998). Recommendation as classification: Using social and content-based information in recommendation. *Proc. of the Fifteenth National Conference on Artificial Intelligence*, July 26–30.

Breese, J.S., Heckerman, D., Kadie, K. (1998). Empirical analysis of predictive algorithms for collaborative filtering. *Proc. 40th Conference on Uncertainty in Artificial Intelligence, UAI'98*, 43–52.

Briki, M., Ben Abdrabbah, S., Ben Amor, N. (2020). Highlighting the users' primary criterion in text mining-based recommendation systems. *OCTA Multi-Conference Proceedings: Information Systems and Economic Intelligence (SIIE)*, 208, February, Tunis, Tunisia [Online]. Available at: https://multiconference-octa.loria.fr/multiconference-program/.

Chu, W.T. and Tsai, Y.L. (2017). A hybrid recommendation system considering visual information for predicting favorite restaurants. *World Wide Web*, 20, 1313–1331.

Ebadi, A. and Krzyzak, A. (2016). A hybrid multi-criteria hotel recommender system using explicit and implicit feedbacks. *Proc. 18th International Conference on Applied Science in Information Systems and Technology (ICASIST)*, Amsterdam, The Netheralnds.

Ganu, G., Elhadad, N., Marian, A. (2009). Beyond the stars: Improving rating predictions using review text contents. *Proc. 12th International Workshop on the Web and Databases, WebDB'09*, June 28.

Hdioud, F., Frikh, B., Ouhbi, B., Khalil, I. (2017). Multi-criteria recommender systems: A survey and a method to learn new user's profile. *International Journal of Mobile Computing and Multimedia Communications*, 8(4), 20–48.

Herlocker, J.L., Konstan, J.A, Terveen, L.G., Riedl, J.T. (2004). Evaluating collaborative filtering recommender systems. *ACM Transactions on Information Systems*, 22(1), 5–53.

Lee, H.H. and Teng, W.G. (2007). Incorporating multi-criteria ratings in recommendation systems. *IEEE International Conference on Information Reuse and Integration*, 273–278.

Li, X., Wang, H., Yan, X. (2015). Accurate recommendation based on opinion mining. *Genetic and Evolutionary Computing*, 329, 399–408.

Lin, K.P., Lai, C.Y., Chen, P.C., Hwang, S.Y. (2015). Personalized hotel recommendation using text mining and mobile browsing tracking. *Proc. IEEE International Conference on Systems, Man, and Cybernetics (SMC)*, 191–196.

Naw, N. and Hlaing, E.E. (2013). Relevant words extraction method for recommendation system. *Bulletin of Electrical Engineering and Informatics*, 2.

Pohekar, S.D. and Ramachandran, M. (2004). Application of multi-criteria decision making to sustainable energy planning – A review. *Renewable and Sustainable Energy Review*, 8, 365–381

Saaty, T.L. (1985). Decision making for leaders. *IEEE Transactions on Systems, Man, and Cybernetics*, 15(3), 450–452.

Sarwar, B., Karypis, G., Konstan, J., Riedl, J. (2001). Item-based collaborative filtering recommendation algorithms. *The 10th International Conference on World Wide Web*, 285–295.

Shardanand, U. and Maes, P. (1995). Social information filtering: Algorithms for automating "word of mouth". *Proc. the Sigchi Conference on Human Factors in Computing Systems*, 210–217.

Soboroff, I. and Nicholas, C. (1999). Combining content and collaborative in text filtering. *IJCAI'99 Workshop: Machine Learning for Information Filtering*, Stockholm, Sweden.

Talib, R., Hanif, M., Ayesha, S., Fatima, F. (2016). Text mining: Techniques, applications and issues. *International Journal of Advanced Computer Science and Applications*, 7(11), 414–418.

Vozalis, E. and Margaritis, K. (2003). Analysis of recommender systems algorithms. *6th Hellenic European Conference on Computer Mathematics & Its Applications*, 732–745.

Yang, C., Yu, X., Liu, Y., Nie, Y., Wang, Y. (2016). Collaborative filtering with weighted opinion aspects. *Neurocomputing*, 210, 185–196.

Yashvardhan, S., Jigar, B., Rachit, M. (2015). A multi-criteria review-based hotel recommendation system. *IEEE International Conference on Computer and Information Technology; Ubiquitous Computing and Communications; Dependable, Autonomic and Secure Computing; Pervasive Intelligence and Computing*, 687–691.

Zheng, Y. (2019). Utility-based multi-criteria recommender systems. *Proc. of the 34th ACM/SIGAPP Symposium on Applied Computing*, 2529–253.

Ziani, A., Azizi, N., Schwab, D., Aldwairi, M., Chekkai, N., Zenakhra, D., Cheriguene, S. (2017). Recommender system through sentiment analysis. *Proc. 2nd International Conference on Automatic Control, Telecommunications and Signals (ICATS)*, December.

Spammer Detection Relying on Reviewer Behavior Features Under Uncertainty

4.1. Introduction

Today, the Internet gives the opportunity to people worldwide to express and share their opinions and attitudes regarding products or services. These opinions called online reviews have become one of the most important sources of information thanks to their availability and visibility. They are increasingly used by both consumers and organizations. Positive reviews usually attract new customers and bring financial gain. However, negative ones damage the e-reputation of different businesses, which leads to a loss. Reviewing has changed the face of marketing in this new area. Due to their important impact, companies invest money to over-qualify their product to gain insights into readers' preferences. To do this, they rely on spammers to usually post deceptive reviews: positive ones to attract new customers and negative ones to damage competitors' e-reputation. These fraudulent activities are extremely harmful to both companies and readers. Hence, detecting and analyzing the opinion spam becomes pivotal for saving e-commerce and to ensure trustworthiness and equitable competition between different products and services. Therefore, different researchers have given considerable attention to this challenging problem. In fact, several studies (Deng and Chen 2014; Ong et al. 2014; Rayana and Akoglu 2015; Heydari et al. 2016; Fontanarava et al. 2017) have been devoted to

Chapter written by Malika Ben Khalifa, Zied Elouedi and Eric Lefèvre.

developing methods capable of spotting fake reviews and stopping these misleading actions. These approaches can be classified into three global categories: spam review detection based on the review content and linguistic features, group spammer detection based on relational indicators and spammer detection.

Since spammers are mainly responsible for the appearance of deceptive reviews, spotting them is surely one of the most essential tasks in this field. Several approaches addressed this problem (Heydari et al. 2016) and succeeded in achieving significant results. Spammer detection techniques can be divided into two global categories: graph-based method and behavioral indicator-based methods.

One of the first studies that relies on graph representation to detect fake reviews was proposed in Wang et al. (2011). This method attempted to spot fake reviewers and reviews of online stores. This approach is based on a graph model composed of three types of nodes, which are reviewers, reviews and stores. The spamming clues are comprised through the interconnections and the relationships between nodes. The detection of these clues is based on the trustworthiness of reviewers, the honesty of reviews and the reliability of stores. Thanks to these three measures, the method generates a ranking list of spam reviews and reviewers. This method was tested on real dataset extracted from resellerratings.com and labeled by human experts and judged. However, the accuracy of this method is limited to 49%. A similar study was proposed by Fayazbakhsh and Sinha (2012) based also on the review graph model. This method generates a suspicion score for each node in the review graph and updates these scores based on the graph connectivity using an iterative algorithm. This method was performed using a dataset labeled through human judgment. Moreover, the third graph-related approach was introduced by Akoglu et al. (2013) as an unsupervised framework. This method relies on a bipartite network composed of reviewers and products. The review can be positive or negative according to the rating. The method assumes that the spammers usually write positive reviews for bad products and negative ones for good quality products. The authors use an iterative propagation algorithm as well as the correlations between nodes and assign a score to each vertex and update it using the loopy belief propagation (LBP). This method offers a list of scores to rank reviewers and products in order to get k clusters. Results were compared to two iterative classifiers, where they have shown performance.

The aspect of the behavior indicators was introduced by Lim et al. (2010) to detect spammers. This method measures spamming behaviors and accords a score to rank reviewers regarding the rating they give. It is essentially based on the assumption that fake reviewers target specific products and that their reviews rating deviates from the average rating associated with these products. Authors assume that this method achieved significant results. Another method proposed in Savage et al. (2015) is also based on the rating behavior of each reviewer. It focuses on the gap between the majority of the given rating and each reviewer's rating. This method uses the binomial regression to identify spammers. One of the most preferred studies was conducted by Fei et al. (2013), which is essentially based on various spammers' behavioral patterns. Since the spammers and the genuine reviewers display distinct behaviors, the proposed method models each reviewer's spamicity while observing their actions. It was formulated as an unsupervised clustering problem in a Bayesian framework. The proposed technique was tested on data from Amazon and proves its effectiveness. Moreover, Fei et al. (2013) proposed a method to detect the burst pattern in reviews given to some specific products or services. This approach generates five new spammer behavioral indicators to enhance review spammer detection. The authors used the Markov random fields to model the reviewers in burst and a hidden node to model the reviewer spamicity. Then, they rely on the loopy belief propagation framework to spot spammers. This method achieves 83.7% of precision thanks to the spammers' behavioral indicators. Since then, behavioral indicators have become an important basis for the spammer detection task. These indicators are used in several recent studies (Liu et al. 2017). Nevertheless, we believe that the information or the reviewers' history can be imprecise or uncertain. Also, the deceptive behavior of users might be due to some coincidence, which makes the spammer detection issue full of uncertainty. For these reasons, ignoring such uncertainty may deeply affect the quality of the detection. To manage these concerns, we propose a novel method that aims to classify reviewers into spammer and genuine ones based on K-nearest neighbors' algorithm within the belief function theory to deal with the uncertainty involved by the spammer behavioral indicators, which are considered as features. It is known as one of the richest theories in dealing with all the levels of imperfection from total ignorance to full certainty. In addition, it allows us to manage different pieces of evidence, not only to combine them but also to make decisions while facing imprecision and imperfections. This theory proves its robustness in this field through our previous methods, which achieve significant results (Ben Khalifa et al. 2018a, 2018b, 2019a, 2019b, 2019c, 2020). Furthermore, the use of the Evidential K-NN (Denoeux 1995) has been based on its robustness in real-world classification problems under uncertainty. We seek to involve imprecision in spammers' behavioral indicators, which are considered the fundamental interest in our approach since they are used as features for the evidential K-NN. In such a way, our method distinguishes between spammers and innocent reviewers while offering an uncertain output, which is the spamicity degree related to each user.

This chapter is structured as follows: in the second section, we present the basic concepts of the belief function theory and the evidential K-nearest neighbors, then we elucidate the proposed method in section 4.3. Section 4.4 presents the experimental results, and we finish with a conclusion and some future work.

4.2. Background

In this section, we elucidate the fundamentals of the belief function theory as well as the evidential K-nearest neighbors classifier.

4.2.1. *The belief function theory*

The belief function theory, called also Dempster Shafer theory, is one of the powerful theories that handles uncertainty in different tasks. It was introduced by Shafer (1976) as a model to manage beliefs.

4.2.1.1. *Basic concepts*

In this theory, a given problem is represented by a finite and exhaustive set of different events called the frame of discernment Ω. 2^{Ω} is the power set of Ω that includes all possible hypotheses, and it is defined by: $2^{\Omega} = \{A : A \subseteq \Omega\}$.

A basic belief assignment (bba) defined as a function from 2^{Ω} to $[0,1]$ that represents the degree of belief given to an element A such that:

$$\sum_{A \subseteq \Omega} m^{\Omega}(A) = 1$$

A focal element A is a set of hypotheses with positive mass value $m^{\Omega}(A) > 0$.

Several types of bbas have been proposed (Smets 1992) in order to model special situations of uncertainty. Here, we present some special cases of bbas:

– the certain bba represents the state of total certainty, and it is defined as follows:

$$m^{\Omega}(\{ \omega_i \}) = 1, \text{where } \omega_i \in \Omega;$$

– simple support function: in this case, the bba focal elements are $\{A, \Omega\}$. A simple support function is defined as the following equation:

$$m^{\Omega}(X) = \begin{cases} \omega & if\ X = \Omega \\ 1 - \omega & if\ X = A\ for\ some\ A \subset \Omega \\ 0 & Otherwise \end{cases}$$

where A is the focus and $\omega \in [0,1]$.

4.2.1.2. Combination rules

Various combination rules have been suggested in the framework of belief functions to aggregate a set of bba's provided by pieces of evidence from different experts. Let m_1^{Ω} and m_2^{Ω} be two bbas modeling two distinct sources of information defined on the same frame of discernment Ω. In the following, we elucidate the combination rules related to our approach.

1) Conjunctive rule: it was settled in Smets (1992), denoted by ⓞ and defined as:

$$m_1^{\Omega} \ ⓞ \ m_2^{\Omega}(A) = \sum_{B \cap C = A} m_1^{\Omega}(B)\, m_2^{\Omega}(C)$$

2) Dempster's rule of combination: this rule is the normalized version of the conjunctive rule (Dempster 1967). It is denoted by and defined as:

$$m_1^{\Omega} \oplus m_2^{\Omega}(A)$$
$$= \begin{cases} \dfrac{m_1^{\Omega} \ ⓞ \ m_2^{\Omega}(A)}{1 - m_1^{\Omega} \ ⓞ \ m_2^{\Omega}(\emptyset)} & if\ A \neq \emptyset, \forall A \subseteq \Omega \\ 0 & otherwise \end{cases}$$

4.2.1.3. *Decision process*

The belief function framework provides numerous solutions to make a decision. Within the transferable belief model (TBM) (Smets 1998), the decision process is performed at the pignistic level where bbas are transformed into the pignistic probabilities denoted by *BetP* and defined as:

$$BetP(B) = \sum_{A \subseteq \Omega} \frac{|A \cap B|}{|A|} \frac{m^{\Omega}(A)}{(1 - m^{\Omega}(\emptyset))} \quad \forall\, B \in \Omega$$

4.2.2. *Evidential K-nearest neighbors*

The evidential K-nearest neighbors (EKNN) (Denoeux 1995) is one of the well-known classification methods based in the belief function framework. It performs the classification over the basic crisp KNN method thanks to its ability to offer a credal classification of the different objects. This credal partition provides a richer information content of the classifier's output.

4.2.2.1. *Notations*

$-\Omega = \{C_1, C_2, ..., C_N\}$: the frame of discernment containing the N possible classes of the problem.

$- X_i = \{X_1, X_2, ..., X_m\}$: the object X_i belonging to the set of m distinct instances in the problem.

$-$ A new instance X to be classified.

$- N_K(X)$: the set of the K-nearest neighbors of X.

4.2.2.2. *EKNN method*

The main objective of the EKNN is to classify a new object X based on the information given by the training set. A new instance X to be classified must be allocated to one class of the $N_K(X)$ founded on the selected neighbors. Nevertheless, the knowledge that a neighbor X_i belongs to class C_q may be deemed as a piece of evidence that raises the belief that the object X to be classified belongs to the class C_q. For this reason, the EKNN technique deals with this fact and treats each neighbor as a piece of evidence that support some hypotheses about the class of the pattern X to be classified. In fact, the more the distance between X and X_i reduces, the more the evidence is strong. This evidence can be illustrated by a simple support function with a *bba* such that:

$$m_{\{X,X_i\}}(\{C_q\}) = \alpha_0 \, exp^{(-(\gamma_q^2 d(X,X_i)^2))}$$

$$m_{\{X,X_i\}}(\Omega) = 1 - \alpha_0 \, exp^{(-(\gamma_q^2 d(X,X_i)^2))}$$

where:

– α_0 is a constant that has been fixed at 0.95;

– $d(X, X_i)$ represents the Euclidean distance between the instance to be classified and the other instances in the training set;

– γ_q assigned to each class C_q has been defined as a positive parameter. It represents the inverse of the mean distance between all the training instances belonging to the class C_q.

After the generation of the different bbas by the K-nearest neighbors, they can be combined through the Dempster combination rule as follows:

$$m_X = m_{X,X_1} \oplus \dots \oplus m_{X,X_K}$$

where $\{1, ..., K\}$ is the set including the indexes of the K-nearest neighbors.

4.3. Spammer detection relying on the reviewers' behavioral features

The idea behind our method is to take into account the uncertain aspect in order to improve detecting spammer reviewers. For that, we propose a novel approach based on different spammer indicators and we rely on the evidential K-nearest neighbors, which is a famous classifier under the belief function framework. In the remainder of this section, we will elucidate the different steps of our proposed approach: in the first step, we model and calculate the spammers' indicators through the reviewers' behaviors. In the second step, we present the initialization and learning phase. Finally, we distinguish between spammers and innocent reviewers through the classification phase in which we also offer an uncertain input to report the spamicity degree of each reviewer. Figure 4.1 illustrates our method steps.

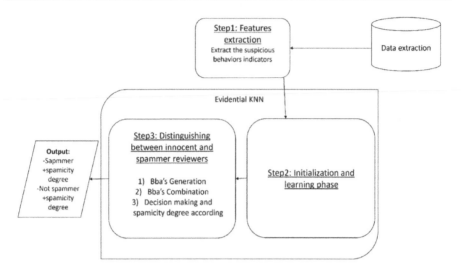

Figure 4.1. *Spammer detection relying on the reviewers' behavioral features*

4.3.1. *Step 1: Features extraction*

As mentioned above, spammer indicators become one of the most powerful tools in the spammer detection field used in several studies. In this section, we suggest controlling the reviewers' behaviors if they are linked with the spamming activities and thus can be used as features to learn the evidential KNN classifier in order to distinguish between the two classes: spammer and innocent reviewers. We select the significant features used in previous work (Mukherjee et al. 2013a, 2013b). Here, we detail them in two lists: in the first list, we elucidate the author, and in the second list, we present the review features. To make the equations more comprehensible, we present the different notations in Table 4.1.

4.3.1.1. *Reviewer features*

The values of these features are in the interval [0,1]. The more the value is close to 1, the higher the spamicity degree is indicated.

4.3.1.1.1. Content similarity (CS)

Generally, spammers choose to copy reviews from other similar products because, for them, creating a new review is considered as an action that required time. Hence, we assume that it is very useful to detect the reviews' content similarity (using cosine similarity). From this perspective and in order to pick up the most unpleasing behavior of spammers, we use the maximum similarity:

$$f_{cs}(R_i) = \max_{R_i(r_j), R_i(r_k) \in R_i(T_r)} cosine(R_{i(r_j)}, R_{i(r_k)})$$

where $R_i(r_j)$ and $R_i(r_k)$ are the reviews written by the reviewer R_i and $R_i(T_r)$ represents all the reviews written by the reviewer R_i.

R_i	A reviewer
$R = \{R_1, ..., R_H\}$	Set of reviewers, where H is the total number of reviewers
R	A review
P	A product
Tr	Total number of reviews
$R_i(r)$	Review written by the reviewer R_i
$R_i(T_r)$	Set of all reviews written by the reviewer R_i
Tr(p)	Set of reviews on product or service p
r(p)	Review on product p
$R_i(r(p))$	Review given by the reviewer R_i to the same product p
$R_i(Tr(p))$	Set of all reviews given by the reviewer R_i to the same product p
$R_i(Tr_*(p))$	Set of rating reviews given by the reviewer R_i to the same product p
$L(R_i)$	Last posting date of the review written by the reviewer R_i
$F(R_i)$	First posting date of the review written by the reviewer R_i
A(p)	The date of the product launch
I_i	Spamming indicator
$S_{mean}(p)$	The mean score of a given product p
S(p)	The reviewing score of the reviews given to one product p by the same reviewer R_i.
$\Omega = \{S, \bar{S}\}$	The frame of discernment including the spammer and not spammer class

Table 4.1. *List of notations*

4.3.1.1.2. Maximum number of reviews (MNR)

Creating reviews and posting them successively in one day displays an indication of deviant behavior. This indicator calculates the maximum number of reviews per day (*MaxRev*) for a reviewer normalized by the maximum value for our full data.

$$f_{MNR}(R_i) = \frac{MaxRev(R_i)}{\max\limits_{R_j \in R} MaxRev(R_j)}$$

4.3.1.1.3. Reviewing burstiness (BST)

Although authentic reviewers publish their reviews from their accounts occasionally, opinion spammers represent a non-old-time membership in the site. To this point, it makes us able to take advantage of the accounts' activity in order to capture the spamming behavior. The activity window, which is the dissimilarity between the first and last dates of the review creation, is used as a definition of the reviewing burstiness. Consequently, if the timeframe of a posted review was reasonable, it could mention a typical activity. Nevertheless, posting reviews in a short and nearby burst ($\tau = 28$ days, estimated by Mukherjee et al. (2013a, 2013b)) shows an emergence of spam behavior.

$$f_{BST}(R_i) = \begin{cases} 0, & L(R_i) - F(R_i) > \tau \\ \dfrac{L(R_i) - F(R_i)}{\tau}, & Otherwise \end{cases}$$

where $L(R_i)$ represents the last posting date of the review r given by the reviewer R_i and $F(R_i)$ is the first posting date of the review.

4.3.1.1.4. Ratio of first reviews (RFR)

To take advantage of the reviews, people rely on the first posted reviews. For this reason, spammers tend to create them at an early stage in order to affect the elementary sales. Therefore, spammers believe that managing the first reviews of each product could empower them to govern people's sentiments. For every single author, we calculate the ratio between the first reviews and the total reviews. We mean by the first reviews those posted by the author as the first to evaluate the product:

$$f_{RFR}(R_i) = \frac{|R_i(r_f)|}{|R_i(T_r)|}$$

where $R_i(r_f)$ represents the first reviews of the reviewer R_i.

4.3.1.2. *Review features*

These features have a binary value. If the feature value is equal to 1, then it indicates spamming. If not, it represents non-spamming.

4.3.1.2.1. Duplicate/near-duplicate reviews (DUP)

As far as they want to enhance the ratings, spammers frequently publish multiple reviews. They tend to use a duplicate/near-duplicate kind of preceding reviews about the same product. We could spotlight this activity by calculating the duplicate reviews on the same product. The calculation proceeds as follows:

$$f_{DUP}R_i(r_j) = \begin{cases} 1, & \text{if } cosine\left(R_i(r_j), R_i(r_k)\right) > \beta_1 \ \forall \ r_j, r_k \in R_i(T_r(p)) \\ 0, & Otherwise \end{cases}$$

For a review r_j, each author R_i on a product p acquires as value 1 if it is in analogy (using cosine similarity based on some threshold, $\beta_1 = 0.7$) with another review is estimated in Mukherjee et al. (2013a, 2013b).

4.3.1.2.2. Extreme rating (EXT)

In favor of bumping or boosting a product, spammers often review it while using extreme ratings (1* or 5*). We have a rating scale composed of five stars (*):

$$f_{EXT}(R_i(r))$$
$$= \begin{cases} 1, & \text{if } \left|R_i(Tr * (p)) \in \{1,5\}\right| > \left|R_i(Tr * (p)) \in \{2,3,4\}\right| \\ 0, & Otherwise \end{cases}$$

where $R_i(Tr * (p))$ represents all the reviews (ratings) given by the reviewer R_i to the same product p.

4.3.1.2.3. Rating deviation

Spammers aim to promote or demote some target products or services to this point they generate reviews or rating values according to the situation. In order to deviate the overall rating of a product, they have to contradict the given opinion by posting deceptive ratings strongly deviating the overall mean.

If the rating deviation of a review exceeds a threshold $\beta_2 = 0.63$ estimated in Mukherjee et al. (2013a, 2013b), this feature achieves the value of 1. The maximum deviation is normalized to four on a five-star scale.

$$f_{Dev}(R_i) = \begin{cases} 1, & \dfrac{|S(p) - S_{mean}(p)|}{4} > \beta_2 \\ 0, & Otherwise \end{cases}$$

where $S_{mean}(p)$ represents the mean score of a given product and $S(p)$ represents the score of the reviews given to one product p by the same reviewer R_i.

4.3.1.2.4. Early time frame (ETF)

Since the first review is considered as a meaningful tool to gauge the sentiment of people on a product, spammers set to review at an early level in order to press the spam behavior. The feature below is proposed as a way to detect the spamming characteristic:

$$ETF\,(R_i(r(p))) = \begin{cases} 0, & L(R_i, p) - A(p) > \delta \\ 1 - \dfrac{L(R_i, p) - A(p)}{\delta}, & Otherwise \end{cases}$$

$$f_{ETF}(r) = \begin{cases} 1, & ETF\left(R_i(r(p))\right) > \beta_3 \\ 0, & Otherwise \end{cases}$$

where $L(R_i, p)$ represents the last review posting date by the reviewer R_i on the product p and $A(p)$ is the date of the product launch. The degree of earliness of an author R_i who had reviewed a product p is captured by $ETF\,(R_i(r(p)))$; the threshold symbolizing earliness is about $\delta = 7$ months (estimated in Mukherjee et al. (2013a, 2013b)). According to the presented definition, we cannot consider the last review as an early one if it has been posted beyond 7 months since the product's launch. On the other hand, the display of a review following the launch of the product allows this feature to reach the value of $\beta_3 = 0.69$, which is considered as the threshold mentioning spamming and is estimated in Mukherjee et al. (2013a, 2013b).

4.3.1.2.5. Rating abuse (RA)

To bring up the wrong use generated from the multiple ratings, we adopt the feature of rating abuse (RA). Obtaining multiple rating on a unique product is considered a strange behavior. Despite the fact that this feature is like DUP, it does not focus on the content but rather it targets the rating dimension. As a definition, the rating abuse, the similarity of the donated ratings by an author for a product beyond multiple ratings by the same author blended by the full reviews on this product:

$$RA(R_i, R_i(r(p))) = |R_i(Tr(p))|(1 - \frac{1}{4}\max_r(r,p) - \min_r(r,p)),$$

where $r \in R_i(Tr(p))$

$$f(RA) = \begin{cases} 1, & RA(R_i, R_i(r(p))) > \beta_4 \\ 0, & Otherwise \end{cases}$$

We should calculate the difference between the two extremes (maximum/minimum) on five-star scale rating to catch the coherence of high/low rating and to determine the similarity of multiple star rating. The maximum difference between ratings attains 4 as a normalized constant. Lower values are reached by this feature if, in authentic cases, the multiple ratings change (as a result of a healthy use). $\beta_4 = 2.01$ is considered as the threshold mentioning spamming and is estimated in Mukherjee et al. (2013a, 2013b).

4.3.2. Step 2: Initialization and learning phase

In order to apply the evidential K-NN classifier, we should first assign values to parameters α_0 and γ_0 to be used in the learning phase. We will start by initializing the parameter α_0 and then computing the second parameter γ_ω while exploiting the reviewer-item matrix. As mentioned in the EKNN procedure (Denoeux 1995), α_0 is initialized to 0.95. The value of the parameter α_0 is assigned only one time, while the γ_ω value changes each time according to the class. In order to ensure the γ_ω computation performance, we select the reviewers who belong to the same class and we assign a parameter γ_ω that will be measured as the inverse of the average distance between each pair of reviewers R_i and R_j having the same class ω. This calculation is based on the Euclidean distance denoted $d(R_i, R_j)$ such that:

$$d(R_i, R_j) = \sqrt{\sum_{k=1}^{n}\left(R_i^k - R_j^k\right)^2}$$

where n represents the number of indicators and R_i^k is the value of the kth indicator for the reviewer R_i.

Once the spammer indicators are calculated and the two parameters α_0 and γ_ω have been assigned, we must select a set of reviewers. Then, we compute for each reviewer R_i in the database their distance to the target reviewer R. Given a target reviewer, we have to spot its K-most similar neighbors, by selecting only the K reviewers having the smallest distances values that is calculated using the Euclidean distance and denoted by $d(R_i, R_j)$.

4.3.3. *Step 3: Distinguishing between innocent and spammer reviewers*

In this step, we aim to classify a new reviewer into a spammer or an innocent reviewer. Let $\Omega = \{S, \bar{S}\}$, where S represents the class of the spammer reviewers and $\bar{S}$ includes the class of the genuine reviewers.

4.3.3.1. *The bba generation*

Each reviewer R_i induces a piece of evidence that builds up our belief about the class that they belong to. However, this information does not supply certain knowledge about the class. In the belief function framework, this case is shaped by simple support functions, where only a part of belief is committed to $\omega \in \Omega$, where ω is the label of the reviewer R_i and the rest is assigned to Ω. Thus, we obtain the following *bba*:

$$m_{R,R_i}(\{\omega\}) = \alpha_{R_i}$$

$$m_{R,R_i}(\Omega) = 1 - \alpha_{R_i}$$

where R is the new reviewer and $R_i \subset N_k(R)$ is their similar reviewer, $\alpha_{R_i} = \alpha_0 \exp^{((-\gamma_\omega)^2\, d(R,R_i)^2)}$, and α_0 and γ_ω are two parameters assigned in the initialization phase.

In our case, each neighbor of the new reviewer has two possible hypotheses. It can be similar to a spammer reviewer in which their committed belief is allocated to the spammer class S and the rest to the frame of discernment Ω. In the other case, it can be near to an innocent reviewer where the committed belief is given to the not spammer class $\bar{S}$ and the rest of is assigned to Ω. We treat the K-most similar reviewers as independent sources of evidence where each one is modeled by a basic belief assignment. Hence, K different bbas can be generated for each reviewer.

4.3.3.2. The combination of bbas

After the generation of the bbas for each reviewer $R_i \subset N_k(R)$, we describe how to aggregate these bbas in order to get the final belief about the reviewer classification. Under the belief function framework, such bbas can be combined using the Dempster combination rule. Therefore, the obtained bba represents the evidence of the K-nearest neighbors regarding the class of the reviewer. Hence, this global mass function m is obtained as such:

$$m_R = m_{R,R_1} \oplus m_{R,R_2} \oplus \dots \oplus m_{R,R_k}$$

4.3.3.3. Decision-making

We apply the pignistic probability *BetP* in order to select the membership of the reviewer R_i to one of the classes of Ω and to accord them a spamicity degree. Then, the classification decision is made: either the reviewer is a spammer or not. For this, the possible label that has the greater pignistic probability is selected. Moreover, we assign to each reviewer even they are not a spammer the spamicity degree which consists of the *BetP* value of the spammer class.

4.4. Experimental study

The evaluation in the fake reviews detection problem was always a challenging issue due to the unavailability of true real-world growth data and variability of the features as well as the classification methods used by the different related work, which can lead to unsafe comparison in this field.

4.4.1. *Evaluation protocol*

4.4.1.1. *Data description*

In order to test our method performance, we use two datasets collected from yelp.com. These datasets represent the more complete, largest, the more diversified and general-purpose labeled datasets that are available today for the spam review detection field. They are labeled through the classification based on the yelp filter, which has been used in various previous works (Mukherjee et al. 2013b; Rayana and Akoglu 2015; Fontanarava et al. 2017; Ben Khalifa et al. 2019b, 2019c) as ground truth in favor of its efficient detection algorithm based on experts' judgment and on various behavioral features. Table 4.2 introduces the datasets content where the percentages indicate the filtered fake reviews (not recommended) and also the spammer reviewers.

The YelpNYC dataset contains reviews of restaurants located in New York City; the Zip dataset is bigger than the YelpNYC datasets, since it includes businesses in various regions of the United States, such as New York, New Jersey, Vermont, Connecticut and Pennsylvania. The strong points of these datasets are:

– the high number of reviews per user, which facilities modeling of the behavioral features of each reviewer;

– the miscellaneous kinds of entities reviewed, i.e. hotels and restaurants;

– above all, the datasets hold just fundamental information, such as the content, label, rating and date of each review, connected to the reviewer who generated them. With regard to considering over-specific information, this allows us to generalize the proposed method to different review sites.

Datasets	Reviews (filtered %)	Reviewers (spammer %)	Services (restaurant or hotel)
YelpZip	608,598 (13.22%)	260,277 (23.91%)	5,044
YelpNYC	359,052 (10.27%)	160,225 (17.79%)	923

Table 4.2. *Datasets description*

4.4.1.2. *Evaluation criteria*

We rely on the following three criteria to evaluate our method: accuracy, precision and recall, which can be defined as the following equations respectively, where *TP*, *TN*, *FP* and *FN* denote true positive, true negative, false positive and false negative respectively:

$$Accuracy = \frac{(TP + TN)}{(TP + TN + FP + FN)}$$

$$Precision = \frac{TP}{(TP + FP)}$$

$$Recall = \frac{TP}{(TP + FN)}$$

4.4.2. *Results and discussion*

Our method relies on the evidential KNN to classify the reviewer into spammers and genuine ones. We compare our method to the support vector machine (SVM) and the Naive Bayes (NB); used by most of spammer detection method (Mukherjee et al. 2013a; Rayana and Akoglu 2015; Liu et al. 2017) in this field. Moreover, we suggest comparing also with our previously proposed uncertain classifier to detect spammers (UCS) in Ben Khalifa et al. (2019a). Table 4.3 reports the different results.

Our method achieves the best performance detection according to accuracy, precision and recall over-passing the baseline classifier. We record at best an accuracy improvement of over 24% in both yelpZip and yelpNYC datasets compared to NB and over 19% compared to SVM. Moreover, the improvement records between our two uncertain methods (Ben Khalifa et al. 2019a) (over 10%) at best shows the importance of the variety of the features used in our proposed approach.

Our method can be used in several fields by different review websites. In fact, these websites must block detected spammers in order to stop the appearance of fake reviews. Furthermore and thanks to our uncertain output which represents the spamicity degree for each reviewer, they can control the behavior of the genuine ones with a high spamicity degree to prevent their tendency to turn into spammers.

Evaluation criteria	Accuracy				Precision				Recall			
Methods	NB	SVM	UCS	Our method	NB	SVM	UCS	Our method	NB	SVM	UCS	Our method
YelpZip	60%	65%	78%	**84%**	57%	66%	76%	**85%**	63%	68%	74%	**86%**
YelpNYC	61%	68%	79%	**85%**	62%	69%	79%	**86%**	61.8%	67.8%	76.7%	**83.6%**

Table 4.3. *Comparative results*

4.5. Conclusion and future work

In this chapter, we tackle the spammer review detection problem and put forward a novel approach that aims to distinguish between spammers and innocent reviewers while taking into account the uncertainty in the different suspicious behavioral indicators. Our method shows its performance in detecting the spammers' review while according a spamicity degree to each reviewer. Our proposed approach can be useful for different review sites in various fields. Moreover, our uncertain input can be used by other methods to model the reliability of each reviewer. In future, we aim to tackle the group spammer aspect in the interest of improving detection in this field.

4.6. References

Akoglu, L., Chandy, R., Faloutsos, C. (2013). Opinion fraud detection in online reviews by network effects. *Proceedings of the Seventh International Conference on Weblogs and Social Media, ICWSM*, 13, 2–11.

Bandakkanavar, R.V., Ramesh, M., Geeta, H. (2014). A survey on detection of reviews using sentiment classification of methods. *IJRITCC*, 2(2), 310–314.

Ben Khalifa, M., Elouedi, Z., Lefèvre, E. (2018a). Fake reviews detection under belief function framework. *Proceedings of the International Conference on Advanced Intelligent System and Informatics (AISI)*, 395–404.

Ben Khalifa, M., Elouedi, Z., Lefèvre, E. (2018b). Multiple criteria fake reviews detection using belief function theory. *The 18th International Conference on Intelligent Systems Design and Applications (ISDA)*, 315–324.

Ben Khalifa, M., Elouedi, Z., Lefèvre, E. (2019a). Fake reviews detection based on both the review and the reviewer features under belief function theory. *The 16th International Conference Applied Computing*, 123–130.

Ben Khalifa, M., Elouedi, Z., Lefèvre, E. (2019b). Spammers detection based on reviewers' behaviors under belief function theory. *The 32nd International Conference on Industrial, Engineering and Other Applications of Applied Intelligent Systems (IEA/AIE)*, 642–653.

Ben Khalifa, M., Elouedi, Z., Lefèvre, E. (2019c). Multiple criteria fake reviews detection based on spammers' indicators within the belief function theory. *The 19th International Conference on Hybrid Intelligent Systems (HIS)*, Bhopal.

Ben Khalifa, M., Elouedi, Z., Lefèvre, E. (2019d). An evidential spammer detection based on the suspicious behaviors' indicators. *The International Multiconference OCTA 2019*, February 6–8.

Ben Khalifa, M., Elouedi, Z., Lefèvre, E. (2020) Evidential group spammers detection. In *Information Processing and Management of Uncertainty in Knowledge-Based Systems*, Lesot. M.J., Vieira, S., Reformat, M., Carvalho, P.J., Wilbik, A., Bouchon-Meunier, B., Yager, R. (eds). Springer, Heidelberg.

Dempster, A.P. (1967). Upper and lower probabilities induced by a multivalued mapping. *Annals of Mathematical Statistics*, 38, 325–339.

Deng, X. and Chen, R. (2014). Sentiment analysis based online restaurants fake reviews hyper detection. *Web Technologies and Applications*, 1–10.

Denoeux, T. (1995). A K-nearest neighbor classification rule based on Dempster- Shafer theory. *IEEE Transactions on Systems, Man, and Cybernetics: Systems*, 25(5), 804–813.

Fayazbakhsh, S. and Sinha, J. (2012). Review spam detection: A network-based approach. Final project report, CSE 590 (Data Mining and Networks).

Fei, G., Mukherjee, A., Liu, B., Hsu, M., Castellanos, M., Ghosh, R. (2013). Exploiting burstiness in reviews for review spammer detection. *Proceedings of the Seventh International Conference on Weblogs and Social Media, ICWSM*, 13, 175–184.

Fontanarava, J., Pasi, G., Viviani, M. (2017). Feature analysis for fake review detection through supervised classification. *Proceedings of the International Conference on Data Science and Advanced Analytics*, 658–666.

Heydari, A., Tavakoli, M., Ismail, Z., Salim, N. (2016). Leveraging quality metrics in voting model based thread retrieval. *World Academy of Science, Engineering and Technology, International Journal of Computer, Electrical, Automation, Control and Information Engineering*, 10(1), 117–123.

Jindal, N. and Liu, B. (2008). Opinion spam and analysis. *Proceedings of the 2008 International Conference on Web Search and Data Mining*, ACM, 219–230.

Jousselme, A.-L., Grenier, D., Bossé, É. (2001). A new distance between two bodies of evidence. *Information Fusion*, 2(2), 91–101.

Lim, P., Nguyen, V., Jindal, N., Liu, B., Lauw, H. (2010). Detecting product review spammers using rating behaviors. *Proceedings of the 19th ACM International Conference on Information and Knowledge Management*, 939–948.

Ling, X. and Rudd, W. (1989). Combining opinions from several experts. *Applied Artificial Intelligence an International Journal*, 3(4), 439–452.

Liu, P., Xu, Z., Ai, J., Wang, F. (2017). Identifying indicators of fake reviews based on spammers behavior features. *IEEE International Conference on Software Quality, Reliability and Security Companion (QRS-C)*, 396–403.

Mukherjee, A., Kumar, A., Liu, B., Wang, J., Hsu, M., Castellanos, M. (2013a). Spotting opinion spammers using behavioral footprints. *Proceedings of the ACM International Conference on Knowledge Discovery and Data Mining*, 632–640.

Mukherjee, A., Venkataraman, V., Liu, B., Glance, N. (2013b). What yelp fake review filter might be doing. *Proceedings of the Seventh International Conference on Weblogs and Social Media*, ICWSM, 409–418.

Ong, T., Mannino, M., Gregg, D. (2014). Linguistic characteristics of shill reviews. *Electronic Commerce Research and Applications*, 13(2), 69–78.

Pan, L., Zhenning, X., Jun, A., Fei, W. (2017). Identifying indicators of fake reviews based on spammer's behavior features. *Proceedings of the IEEE International Conference on Software Quality, Reliability and Security Companion*, QRS-C, 396–403.

Rayana, S. and Akoglu, L. (2015). Collective opinion spam detection; bridging review networks and metadata. *Proceedings of the 21th International Conference on Knowledge Discovery and Data Mining*, ACM SIGKDD, 985–994.

Savage, D., Zhang, X., Yu, X., Chou, P., Wang, Q. (2015). Detection of opinion spam based on anomalous rating deviation. *Expert Systems with Applications*, 42 (22), 8650–8657.

Shafer, G. (1976). *A Mathematical Theory of Evidence*, vol. 1. Princeton University Press, Princeton.

Smets, P. (1990). The combination of evidence in the transferable belief model. *IEEE Transactions on Pattern Analysis and Machine Intelligence*, 12(5), 447–458.

Smets, P. (1992). The transferable belief model for expert judgement and reliability problem. *Reliability Engineering and System Safety*, 38, 59–66.

Smets, P. (1995). The canonical decomposition of a weighted belief. *Proceedings of the Fourteenth International Joint Conference on Artificial Intelligence,* 1896–1901.

Smets, P. (1998). The transferable belief model for quantified belief representation. In *Quantified Representation of Uncertainty and Imprecision,* Smets, P. (ed.). Springer, Dordrecht.

Wang, G., Xie, S., Liu, B., Yu, P.S. (2011). Review graph based online store review spammer detection. *Proceedings of 11th International Conference on Data Mining, ICDM,* 1242–1247.

Social Networking Application, Connections Between Visual Communication Systems and Personal Information on the Web

5.1. Introduction

In this chapter, we contribute to the field of information systems by analyzing the connections between social networking sites (SNS) and artificial communication systems (e.g. visual communication systems). Information systems research is usually interdisciplinary, as it involves social sciences, applied sciences, formal sciences and humanities disciplines. Several Web applications, namely SNS, are currently using artificial visual communication systems to facilitate interactions and knowledge exchange between different users and members worldwide. This may be justified by the emergence of global social developments (Löwstedt 2018), as well as an available international audience. For example, we can note the presence of visual communication systems in the Web applications' user-interface and dictionary of emoticons (Alloing and Pierre 2017).

In the following paragraphs, we will define the SNS, as well as artificial communication systems.

Social Networking Sites: an SNS "is a networked communication platform in which participants 1) have uniquely identifiable profiles [...];

Chapter written by Marilou KORDAHI.

2) can publicly articulate connections that can be viewed and traversed by others; and 3) can consume, produce, and/or interact with streams of user-generated content provided by their connections on the site" (Ellison and Boyd 2013, p. 160) (Fernback 1997; Wellman and Gulia 1999; Wellman 2002; Boyd and Ellison 2007; Raine and Wellman 2012; Waheed et al. 2017; Kordahi 2020). The growth of these connections or social ties (strong or weak ties) can only take place if these participants (individuals or organizations) have become members of the SNS (Granovetter 1983; Raine and Wellman 2012). The information exchange may be done in various ways, such as instant messaging, emailing, voice recording, posting.

We will attempt to design the "SignaComm", the first SNS with an internationally oriented communication system for the protection of personal data on the Web. The SignaComm will be informative (Boyd and Ellison 2007; Ma et al. 2011; Ellison and Boyd 2013), and will execute two functions dynamically and in real time. Firstly, it will translate the member's input text into "signagrams" and deliver the result to another member. Secondly, it will display the history of instant messages in the chat room page. Our SNS would be used to deliver information to be understood and used quickly by its members. We hope that users from any culture or social environment, or with disabilities could use it. The protection of personal data is defined by laws and regulations prohibiting the processing, storage or sharing of certain types of information about individuals without their knowledge and consent (e.g. analyzing user's behavior on a website) (Kennedy and Millard 2016).

Artificial communication systems: a number of artificial communication systems have been developed to improve the management of information, regardless of a specific natural language (e.g. Istotype, Universal Playground) (Neurath 1974; Fitrianie and Rothkrantz 2007; Takasaki and Mori 2007). We are interested in the signage system, an artificial visual communication system with an international vocation where the "signagram" is the writing unit (Kordahi 2013a, 2013b). The signagram's type is figurative as it is created from a direct representation of the object that evokes the object or situation to be represented (Klinkenberg 1996). Each signagram is made of an "external shape" (including the contours) and an "internal shape" (Kordahi 2013a) (see Figure 5.1).

The signage system and signagram will be integrated in the SignaComm, to enable internationally oriented communication.

The goal is to present the preliminary results of work in progress on the creation of the "SignaComm". This SNS would support multilingual communication between users worldwide for the protection of personal data on the Web.

We designed the SignaComm while relying on a theory and two principles: the theory of patterns (Alexander 1977, 1979; Gamma et al. 1995; Kraut and Resnick 2012), as well as the principles of ontologies (Gruber 1993; Noy and McGuinness 2001; Gruber 2008) and signage systems (Kordahi 2013a). At the core of Alexander's theory, a pattern describes the characteristics of a generic solution to a specific problem (e.g. the communication in real time between users worldwide). The theory of patterns allows the reuse and remodeling of patterns to serve as resources for software development and problem solving. According to Alexander (1979, p. 313), "each pattern sits at the centre of a network of connections which connect it to certain other patterns that help to complete it". The network of these relationships between small and large patterns creates the pattern. The ontology describes a structured set of concepts and objects by giving meaning to an information system in a specific area (e.g. the user profile), and allows the construction of relationships between these concepts and objects (Gruber 1993; Noy and McGuinness 2001; Gruber 2008).

The SignaComm could be implemented in the structure of a company's or public organization's information system. Many fields may be interested in this SNS, for example, cybersecurity, serious games, online learning. In our case, we are interested in the field of administrative authorities, namely the National Commission for Informatics and Liberty (in French, *Commission nationale de l'informatique et des libertés (CNIL)*). The CNIL is responsible for monitoring the data protection of professionals and individuals. We will explain the approach followed to develop the SignaComm for the protection of personal data when there may be a breach of privacy rights (e.g. email advertising).

Our work consists of six sections. In section 5.2, we will present previously published works. In section 5.3, we will explain the SNS' characteristics and then design its pattern. In section 5.4, we will design the pattern for the automatic translation of text phrases into signagrams for the protection of personal data. In section 5.5, we will develop and test the prototype application that executes the SignaComm and communicates in visual messages, using the signage system and translation software of key phrases into signagrams. In section 5.6, we will discuss the overall approach and finally conclude our work.

5.2. Related published works

To our best knowledge, research projects addressing both topics, the SNS for multilingual communication and the protection of personal data, are limited. However, research projects are conducted on data protection icons mapped to legal information, SNS combined with instant messaging and translation, and automatic translation of text phrases into signagrams. We will use our studies of these related works to fine-tune this research project and create the SignaComm.

In their recently published work, Rossi and Palmirani (2019) described an approach for the creation and assessment of data protection icon sets, which were designed according to human-centered methods. The icon sets were modeled based on an ontology for the general data protection regulation. The XML syntax was used to make the icons machine-readable and dynamically retrievable. Icon sets were mapped to legal data to improve the understanding of the general data protection regulation.

In their published works, Seme (2003) and Yang and Lin (2010) respectively developed a patent and system to automatically translate and send instant messages between members who communicate in different languages. Members, engaged in a session of instant messages, could send a message in a source language that could be translated automatically and received in a target language. The translation process followed the natural language processing approach.

In 2016, we published works regarding an SNS for crisis communication. The objectives have been to translate a sequence of syntagms into a series of signagrams in real time, and facilitate communication between members around the world. This SNS automatically translated a source text into a target text (e.g. a message from the French language to the signage system) and displayed the results in the SNS. The SNS was based on the principles of the signage system, modular architecture and ontologies.

In 2015, with Baltz, we designed a software to automatically translate an input text into a sequence of signagrams. We relied on the semantic transfer method (Emel et al. 2000) with the linguistic rules and dictionaries for the source language and target communication system. The input was the source text and written in the user's preferred language. The output was the target text and written in the visual communication system, signage.

We rely on our works published in 2013 and 2019 to show an example of signagram representing the syntagm *identify partners and data recipient* (National Commission on Informatics and Liberty 2020) (see Figure 5.1). In Figure 5.1, we demonstrate how to design the signagram, while including the shapes, contours and colors.

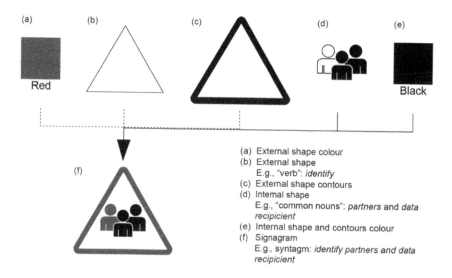

Figure 5.1. *Example of signagram. For a color version of this figure, see www.iste.co.uk/sidhom/systems.zip*

5.3. Pattern for the SignaComm, first approach

We rely on the writings of Alexander (1977, 1979) to create a pattern for the SignaComm (Kraut and Resnick 2012). This pattern is a first approach. The section consists of two main paragraphs, the descriptions of the SNS' context and design.

5.3.1. *SignaComm's context*

In general terms, the SNS interface follows the universal design principles of simplicity, flexibility and accessibility of use (Spiliotopoulos et al. 2013; Chandler 2017). In addition, an SNS interface is graphical and contextual (Morin et al. 2012; Jain et al. 2013). Its graphical nature is based on a template that meets the already defined and precise rules to ensure homogeneous and uniform results. These rules are the following: a simple and figurative content, uniqueness of graphical representations and uniqueness of color contents (Kordahi 2013a). As for the dictionary of emoticons, it contains emotion symbols that are used worldwide (Alloing and Pierre 2017).

So far, we have not found published works regarding the standardization of visual communication systems for SNS. A number of companies have developed their own communication systems to integrate it into the SNS interface (e.g. the graphical user interface of WhatsApp) and new technology tools (e.g. the graphical user interface of Apple iPhone). The company's (or organization's) aim may be to intuitively guide users in their actions in entirely different and various contexts (Jain et al. 2013). Each company (or organization) chooses to adapt the charter of its visual communication system and its corresponding tools (e.g. SNS and new applications) according to the targeted countries. This adaptation approach may include the countries' laws, cultures, customs and traditions. Social media applications (the graphical user interface and emoticon dictionaries included) are essentially altered for two reasons. Firstly, to be compatible with international standards and regulations defined by every country's government, and secondly, to meet the universal design principles depending on users' cultures.

To fulfil its objective, the SignaComm for the protection of personal data should use the signage system (Kordahi 2013a), as well as the graphical and contextual interface (Morin et al. 2012; Jain et al. 2013). The latter should meet the universal design principles (Spiliotopoulos et al. 2013; Chandler 2017). We have chosen both criteria to ensure the SignaComm's perception and spontaneous understanding worldwide. Furthermore, the SignaComm's development is in relation to three main concepts: the signage system (Kordahi 2013a), new SNS technology tools (Boyd and Ellison 2007; Ellison and Boyd 2013) and the user's adaptation. This interrelation makes the SignaComm dependent on these social, technical and human environments.

For now, we will not include the emotional aspect as this SNS is an informative one.

5.3.2. *SignaComm's pattern*

The pattern for the SignaComm holds a network of connections between large and small patterns. In this work, we will present 12 large and two small patterns (in total 14 patterns) (Alexander 1977; Gamma et al. 1995). The description is divided into several stages. A diagram will follow the explanations (see Figure 5.2).

We start with pattern 1 (larger environments), which refers to many environments influencing the growth of SNS, such as national legal systems, Internet governance, Internet security threats, information and communication technologies, architecture design, as well as socio-cultural environments (Boyd and Ellison 2007; Ellison and Boyd 2013). These larger environments belong to the social sciences, applied sciences, formal sciences and humanities disciplines.

Pattern 2 (virtual communities environment) is contained inside pattern 1. A virtual community is an information system with a social network of members (Wellman 2002). The interactions between users can be nominative (e.g. between two specific individuals) or grouped (e.g. between a user and a group). Virtual community users can belong to various geographic locations, cultures, age groups and social groups (Wellman 2002). A virtual community's aim is to allow a group of users to communicate randomly and worldwide (Fernback 1997; Wellman and Gulia 1999) (e.g. exchange learning resources, play). Pattern 2 holds pattern 10 (SignaComm community). Patterns 1 and 2 will influence pattern 10's development and growth (Boyd and Ellison 2007; Raine and Wellman 2012; Ellison and Boyd 2013).

Pattern 10 is designed based on the SNS' characteristics (Boyd and Ellison 2007; Raine and Wellman 2012; Ellison and Boyd 2013). It contains and describes the SignaComm functionalities (pattern 11), information technology administration (IT administration) (pattern 40) and interface (pattern 100). As pattern 10 is contained inside patterns 1 and 2, it interacts with both of them (Boyd and Ellison 2007; Raine and Wellman 2012; Ellison and Boyd 2013).

Pattern 11 has various functionalities, listed as follows: the automatic translation of syntagms into signagrams (pattern 22), signage and signagrams models (pattern 23), natural languages and linguistic rules (pattern 24), dictionary (pattern 25), ontology (pattern 26), user profile characteristics and members' list (patterns 20 and 21), activities (pattern 30) and privacy (pattern 31) (Boyd and Ellison 2007; Raine and Wellman 2012; Ellison and Boyd 2013; Kordahi 2013a). Every functionality has its own programming functions. It is activated instantly according to member's or user's requests. Some functionalities are dynamically synchronized to be able to respond to member's or user's requests (Boyd and Ellison 2007; Raine and Wellman 2012; Ellison and Boyd 2013). For example, pattern 22 is synchronized with patterns 23, 24, 25, 26 and 30.

The SignaComm community (pattern 10) requires that all functionalities (pattern 11) execute their tasks to ensure the smooth running of the SNS. Pattern 12 (boundaries of SignaComm's functionalities) establishes boundaries to each functionality, allowing it to perform its assigned tasks. It avoids the overlap with other functionalities.

Pattern 20 relates to user's profile characteristics (Cardon 2008; Proulx 2012). The SignaComm community encourages the diversity of members in order to enrich its growth (Wellman 2002). Therefore, the growth of the SNS also depends on a well-balanced and represented community of members. This community would be able to support the interactions (pattern 30) between its members. For example, the interactions would help a member to solve a situation (Wellman 2002).

Pattern 20 has links with pattern 21 (members list). The latter pattern specifies the SignaComm target audience (e.g. the bidirectional relationships to define a reciprocated link between two members) (Bouraga et al. 2014). Members would belong to different cultures and social classes, as well as different age groups (Wellman 2002).

Pattern 22 (automatic translation) is a central functionality as it receives members' translation requests from the SNS's interface (pattern 100) and facilitates the communication between SignaComm members (pattern 21). Pattern 22 relies on the signage and signagrams, natural languages and linguistic rules, ontology, as well as dictionary to translate members' requests (pattern 30).

Pattern 30 (activities) is mostly linked to patterns 20, 21, 22, 31, 40 and 100 to create nodes of activities, thus allowing members or groups to engage in various ways (Raine and Wellman 2012). These activities may include invitations to join the SignaComm, instant messaging and geolocation of members with their approval. Here, members have the opportunity to make acquaintances and connections, as well as to chat with members and groups of their choice. Depending on the proximity of members, some ties are strong while others are weak (Granovetter 1983; Ellison and Boyd 2013). These activities are displayed in the SNS's interface (pattern 100), and generated data are stored in the secured database (pattern 40).

Pattern 31 (privacy) is mainly for patterns 2, 20, 21, 22, 30 and 40. This pattern allows every member to set their data sharing options with the IT administration, members and SNS environment (Gross and Acquisti 2005; Cardon 2008; Kennedy and Millard 2016; Proulx 2012; Bouraga et al. 2014; Waheed et al. 2017). It takes every member's wish to accept or forbid the sharing or storage of personal information into consideration. We provide the following example: a member chooses not to publicly display their profile and then not to share their geographical position with the SignaComm and its environment. Pattern 10 (SignaComm community) must respect every member's choice (Kennedy and Millard 2016).

Pattern 40 (IT administration) is connected to patterns 11 (SignaComm functionalities) and 100 (interface). To make the interface and functionalities real, it is necessary to set up an IT administration. The latter manages the database and security, modifies the SNS, analyzes the generated information and answers to members' requests.

Pattern 50 (network of links and ties) creates and manages the network of relationships between all the patterns (Boyd and Ellison 2007; Raine and Wellman 2012; Ellison and Boyd 2013). It allows the information to circulate instantly and correctly in the SignaComm community.

Pattern 100 (interface) gives an overview of the SignaComm interface, with an emphasis on the space of exchange between SignaComm members. The SNS' functionalities and IT administration contribute to its design (Morin et al. 2012; Jain et al. 2013). It includes the universal design principles (Spiliotopoulos et al. 2013; Chandler 2017).

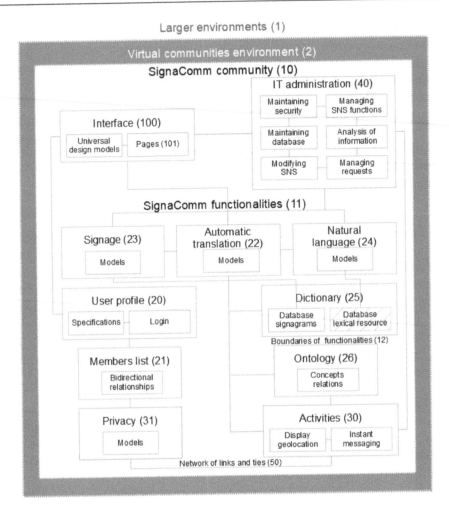

Figure 5.2. *Diagram for the SignaComm pattern*

Pattern 101 (pages) is the continuation of pattern 100. The SignaComm's architecture includes the design of web pages and signagram buttons (Kordahi 2016). The SignaComm is created with a reduced number of pages, such as the registration, members and chat room pages. This design is followed to quickly access information, provide flexibility in use and initiate intuitive interactions (Jain et al. 2013; Spiliotopoulos et al. 2013; Kordahi

2016; Chandler 2017). The navigation between these pages is done through the signagram buttons.

Figure 5.2 shows an overall view of the SignaComm's pattern. It includes the 14 patterns. We show the main links between the patterns to simplify the diagram's representation.

5.4. From text phrases to signagrams for the protection of personal data

Once we have designed the SignaComm pattern, we start developing pattern 22 (automatic translation). As a reminder, the latter is a central functionality to achieve the SignaComm's objective. We rely on Emele et al. (2000)'s works and ours (Kordahi 2013a, 2013b) to accomplish this task. We will explain the methodology of work for developing both the software and dictionary for the protection of personal data.

5.4.1. *Automatic translation*

We analyze the situation where a SignaComm member uses the application to translate a sequence of syntagms (or text phrases) into a series of signagrams in real time, and engage in an informative conversation with a member or group of members. We present information to be quickly understood by members, to prevent some manipulation of personal data without their knowledge or permission and regardless of the computing device used (Kennedy and Millard 2016; National Commission on Informatics and Liberty 2020) (e.g. the portability of data).

The machine translation prototype and its results are constantly assessed according to the knowledge produced. The latter is acquired during the machine translation process and while fulfilling members' requests. Following the automatic translation of key phrases into signagrams, we wish to know how members have understood the translated messages and which translations are useful to them (Suojanen et al. 2014). In other words, a member, who has viewed the translation result, has the choice of editing it. This option allows them to choose the translation which meets their expectations. It is also an invitation to contribute to the machine translation process, in order to provide more precision to the requests made. The member can then confirm their choice or restart the translation request (see Figure 5.3).

While developing this machine translation prototype, we face one main difficulty, namely the non-figurative legal corpus. The suggested solutions are, on the one hand, to segment and analyze a thematic text and, on the other hand, to only translate the syntagms related to the case (Kordahi 2015; Kordahi and Baltz 2015).

Here, for this machine translation, the expressions' exactness is necessary to be able to break down their relations with other encompassing units. This would help by decreasing blunders and uncleanness in the translation process (Bar-Hillel 2003; McShane et al. 2005; Kordahi 2015; Kordahi and Baltz 2015). Consequently, we use the National Commission for Informatics and Liberty portal's thematic text that presents reliable and relevant information.

Our model is composed of the ontology for the protection of personal data (Palmirani *et al.* 2018a, 2018b), the construction of a dictionary of signagrams also related to the protection of personal data (Takasaki 2006; Holtz et al. 2010; Kordahi 2013b) and the adaptation of the function translating text phrases into signagrams (Emel et al. 2000; Seme 2003; Kordahi 2015; Kordahi and Baltz 2015). We use the Natural Language Toolkit in Python (Bird et al. 2009). It allows connections with the ontology and dictionary, in addition to other functionalities (e.g. text processing, semantic reasoning).

We are particularly interested in the works of Palmirani et al. (2018a, 2018b), as their ontology is based on the application of the general data protection regulation. The accuracy, flexibility and reliability of this ontology are well in line with our work objective. Therefore, it is appropriate to integrate it in the project.

Figure 5.3 shows an example of the automatic translation of text phrases into signagrams. The user has the option to edit the translation result, where other signagrams will be shown.

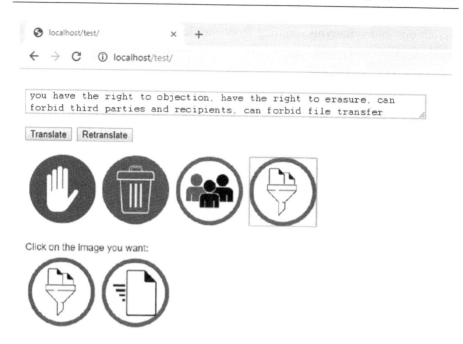

you have the right to objection, have the right to erasure, can forbid third parties and recipients, can forbid file transfer

Translate   Retranslate

Click on the image you want:

Figure 5.3. *Example of a machine translation result. For a color version of this figure, see www.iste.co.uk/sidhom/systems.zip*

5.4.2. *Dictionary of signagrams*

To our knowledge, published works related to the dictionary of signagrams for the protection of personal data are limited. We rely on the works of Takasaki (2006) and Kordahi (2013b) to design and develop this first dictionary, which is specialized. It provides information on signagrams to improve their understanding by any user.

The dictionary's design is based on the correspondence of vector signagrams to homologous semantic-based concepts. We program a mapping between two resources. The first semiotic graphical resource contains signagrams' external shapes (including the contours) and internal shapes (Kordahi 2013a). The external and internal shapes, coming from that graphical resource, are stored in the dictionary. The second resource is a semantic lexical one (e.g. the WordNet (Miller 1998)). The latter contains the concepts with their definitions and synonyms in English. The words, definitions and synonyms, coming from that lexical resource, are contained inside the dictionary (see Figure 5.3).

We create 50 signagrams based on the works of the United Nations Economic Commission for Europe (2006), Holtz et al. (2010) and Rossi and Palmirani (2019), as well as the "Fotolia" international image bank. The latter holds a large collection of images and symbols used globally. The signagrams' colors and shapes follow the international charter road signs (UNECE 2006).

5.5. SignaComm's first technical test

For now, we have designed and programmed a prototype of the SNS. It is implemented in the Elgg platform and hosted on local and private servers.

The SignaComm is written with the Python, PHP and Javascript programming languages to enable queries to be performed from a web page. In this section, we choose to explain the four main patterns that are dynamically connected (see section 5.3) (Alexander 1977, 1979; Gamma et al. 1995; Kraut and Resnick 2012). These patterns are the following: the interface (patterns 100 and 101), user profile (patterns 20 and 21), automatic translation of syntagms into signagrams (patterns 22–25) and activities (pattern 30).

5.5.1. *Interface pattern*

The SignaComm's interface is used to display two sorts of information: the resulting information from an exchange between SNS members and interactions between the SNS system and its members. We provide the following examples, which include sending and receiving instant messages, displaying automatic translation of written texts into a sequence of signagrams and viewing a member's profile (see section 5.3, Figure 5.2).

The graphical user-interface is made up of a set of HTML web pages. It consists of a main interface and secondary one. The main interface is used to display the web pages' content. The secondary interface is the navigation bar. It enables the browsing between the various pages (see section 5.3, Figure 5.4).

5.5.2. *User profile pattern*

We strive to respond to the ethical principles of SNS (Kennedy and Millard 2016; Ellison and Boyd 2013). We mainly focus on the member's professional information (e.g. the profession, name and surname, profile display mode (private or public)). We respect information integrity and analyze the context in which it was saved in the SignaComm database. This information must not be distributed to third parties, before obtaining the owner's consent (Kennedy and Millard 2016).

The user profile pattern performs three essential tasks. These are the registration of a user, invitation of a user and geolocation of members (see section 5.3, Figure 5.2). The first task allows a user to register and login to the SignaComm, which is a condition to use this SNS. The registration is done by submitting a user-name and password, as well as some information regarding the user (e.g. choosing to share their information (Cardon 2008; Morin et al. 2012; Proulx 2012) and geographical position with the SNS) (see Figure 5.4). The sign-in is done by submitting the member's user-name and previously saved password. The second function allows a SignaComm member to invite another user by sending an electronic invitation (e.g. instant message) while using the other patterns (e.g. pattern 21). Pattern 20 is connected to a geolocation process to make it possible to perform the third task. The latter task automatically suggests a language of conversation (Yang and Lin 2010).

This pattern comprises an application page and a PHP function. The HTML application page collects the user's registration information, including the name, physical address, email and address. The collected information is sent to the PHP function.

5.5.3. *Machine translation pattern*

We rely on our works developed in sections 5.3 and 5.4 to implement the machine translation in the SignaComm structure. Once implemented in the SignaComm, the translation pattern runs three consecutive tasks that are stored in this SNS database. The chat room page (written in HTML format) can receive the member's input text. A first request transmits the input text to be automatically translated into vector signagrams. A second request displays the machine translation result in the same HTML page. And a third request waits for the member's action to send the translated message to the activity pattern, edit the translation results or reset the automatic translation process (Seme 2003) (see Figure 5.4).

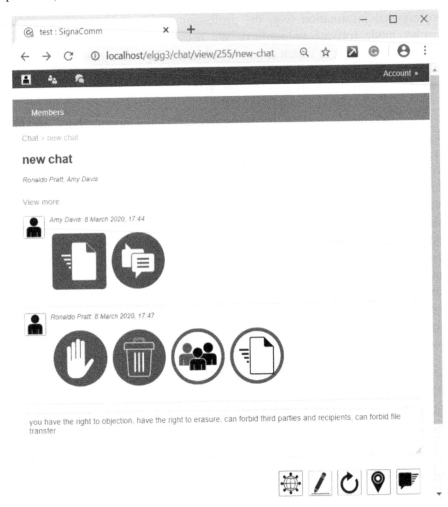

Figure 5.4. *Example of the chat room page. For a color version of this figure, see www.iste.co.uk/sidhom/systems.zip*

Figure 5.4 shows an example of the SignaComm and the translation results. Here, the reported digital identities are simulated using fake profiles. Member 1 writes an input text (*we wish to transfer files and request advice*), activates its automatic translation and then sends the resulting translation to a corresponding member 2. Member 2 replies to member 1 by writing, translating and sending a message. The signagrams' reading direction is from left to right and top to bottom (Neurath 1974). The result of Figure 5.4 is comparable to Figure 5.3.

5.5.4. *Activity pattern*

The pattern of activities performs two simultaneous and programmed tasks that are saved in the SignaComm. Chat histories are saved in the database's tables (see section 5.3, Figure 5.2). Through the server, the translation pattern receives requests from a member in the form of packets compliant with a common Internet protocol (e.g. the HyperText Transfer Protocol (HTTP) POST packets). These packets contain the translated information (the message is translated before delivery). The second task displays the instant messaging exchange between members on the chat room page (Yang and Lin 2010) (see Figure 5.4).

5.6. Discussion and conclusion

While creating the SignaComm, with an internationally oriented communication system for the protection of personal data, we overcame at least one difficulty. To protect personal information on the Web, information accuracy, reliability, flexibility and speed of transmission are needed to assist individuals. We have formed the SignaComm of interrelated patterns. This interrelation has allowed us to synchronize the information exchange (Kordahi 2020).

The obtained results demonstrate that the SignaComm is functioning correctly. In real time and instantly, a sequence of text phrases is translated into a series of signagrams in order to send the results to members. Members can create their own network of contacts by inviting users of their choice. The geolocation process also identifies the member's preferred language.

Moreover, since the SignaComm for the protection of personal data is a first and new prototype, we recommend preparing users for its use with the aim of optimizing its performance. This preparation should include detailed explanations regarding the SNS: the purpose, usefulness of its use, interface functionalities and signage system. This preparation could be done in various ways, such as through a demonstration video, detailed guide and questions and answer (Q&A) forum. Online help would be interesting to design and implement in the SignaComm context. This would explain the SNS' social utility and the meaning of every signagram. Its use may be punctual, used to understand the meaning of a specific signagram or to search for a specific functionality.

In the near future, we would like to analyze and test the SignaComm with other writing systems, for instance Chinese. A collaboration with researchers specialized in linguistics and computer science will be required to understand the Chinese writing system, write the corresponding algorithms and produce exact results. Furthermore, we wish to improve this first prototype. We will place the SignaComm in other fields and contexts to make it more robust and reliable, like the online learning field. In this context, on the one hand, we will analyze the digital identity of different SignaComm users/members, as well as the visibility models (Cardon 2008; Turkle 2011). On the other hand, we will conduct qualitative and quantitative studies on the user's behavior while using the SNS. This study will allow us to evaluate, measure and improve the time required to understand a visual message. Finally, the design of signagrams for personal information on the Web could be part of a research project, in order to further deepen studies and conduct empirical tests (Rossi and Palmirani 2019). The design and verification of signagrams will be made by participants who belong to different fields of specialization to ensure consistent message communication and understanding.

5.7. Acknowledgment

The authors would like to thank Mohammad Haj Hussein, computer and communications engineer, for his valuable help while programming the machine translation prototype.

5.8. References

Alexander, C. (1977). *A Pattern Language: Towns, Buildings, Construction*. Oxford University Press, New York.

Alexander, C. (1979). *The Timeless Way of Building (Vol. 1)*. Oxford University Press, New York.

Alloing, C. and Pierre, J. (2017). *Le web affectif : une économie numérique des émotions*. INA Editions, Paris.

Bar-Hillel, Y. (2003). The present status of automatic translation of languages. *Readings in Machine Translation*, 45–77.

Bird, S., Klein, E., Loper, E. (2009). *Natural Language Processing with Python: Analyzing Text with the Natural Language Toolkit*. O'Reilly Media Inc, California.

Bouraga, S., Jureta, I., Faulkner, S. (2014). Requirements engineering patterns for the modeling of online social networks features. *4th International Workshop on Requirements Patterns*, IEEE, 33–38.

Boyd, D.M. and Ellison, N.B. (2007). Social network sites: Definition, history, and scholarship. *Journal of Computer-Mediated Communication*, 13(1), 210–230.

Cardon, D. (2008). The design of visibility. *Réseaux*, 6, 93–137.

Chandler, D. (2017). *Semiotics: The Basics*. Taylor & Francis, London.

Ellison, N.B. and Boyd, D. (2013). Sociality through social network sites. *The Oxford Handbook of Internet Studies*, 151–172.

Emele, M.C., Dorna, M., Lüdeling, A., Zinsmeister, H., Rohrer, C. (2000). Semantic-based transfer. *Verbmobil: Foundations of Speech-to-Speech Translation*, 359–376.

Fernback, J. (1997). The individual within the collective: Virtual ideology and the realization of collective principles. In *Virtual Culture: Identity and Communication in Cybersociety*, Jones, S. (ed.). SAGE Publications, London.

Fitrianie, S. and Rothkrantz, L.J. (2007). A visual communication language for crisis management. *International Journal of Intelligent Control and Systems (Special Issue of Distributed Intelligent Systems)*, 12(2), 208–216.

Fotolia (2020). Fotolia [Online]. Available at: https://www.fotolia.com.

Gamma, E., Helm, R., Johnson, R., Vlissides, J. (1995). *Design Patterns: Elements of Reusable Object-Oriented Software*. Pearson Education India, Delhi.

Granovetter, M. (1983). The strength of weak ties: A network theory revisited. *Sociological Theory*, 1, 201–233.

Gross, R. and Acquisti, A. (2005). Information revelation and privacy in online social networks. *Proceedings of the 2005 ACM Workshop on Privacy in the Electronic Society*, 71–80.

Gruber, T. (1993). A translation approach to portable ontology specifications. *Knowledge Acquisition*, 5(2), 199–221.

Gruber, T. (2008). Collective knowledge systems: Where the social web meets the semantic web. *Journal of Web Semantics*, 6(1), 4–13.

Holtz, L.E., Nocun, K., Hansen, M. (2010). Towards displaying privacy information with icons. In *IFIP PrimeLife International Summer School on Privacy and Identity Management for Life*, Fischer-Hübner, S., Duquenoy, P., Hansen, M., Leenes, R., Zhang, G. (eds). Springer, Berlin, Heidelberg.

Jain, R., Bose, J., Arif, T. (2013). Contextual adaptive user interface for Android devices. *Annual IEEE India Conference (INDICON)*, 1–5.

Kennedy, E. and Millard, C. (2016). Data security and multi-factor authentication: Analysis of requirements under EU law and in selected EU Member States. *Computer Law & Security Review*, 32(1), 91–110.

Klinkenberg, J.M. (1996). *Précis de sémiotique générale*. De Boeck Université, Brussels.

Kordahi, M. (2013a). Signage as a new artificial communication system. *Canadian Journal of Information and Library Science*, 37(4), 237–252.

Kordahi, M. (2013b). SignaNet : premier dictionnaire électronique spécialisé de la signalétique. *Pratiques et usages numériques H2PTM'2013*, 385–387.

Kordahi, M. (2015) Automatic translation of text phrases into vector images for crisis communication. *6th International Conference on Information Systems and Economic Intelligence (SIIE)*, IEEE, 119–124.

Kordahi, M. (2016). Réseau social numérique et images vectorielles : introduction à une communication à vocation internationale. *Recherches en communication*, 42(42), 233–251.

Kordahi, M. (2019). La signalétique comme système de communication internationale, la protection des informations personnelles sur le Web. *International Association for Media and Communication Research* [Online]. Available at: https://www.academia.edu/40275075/iamcr_2019_-_communication_ technology_and_human_dignity_disputed_rights_contested_truths [Accessed 9 April 2020].

Kordahi, M. (2020). Social networking application, visual communication system for the protection of personal information. *The International Multiconference OCTA 2019 Proceedings*, IEEE, Tunis, 6–8 February [Online]. Available at: https://multiconference-octa.loria.fr/multiconference-program/.

Kordahi, M. and Baltz, C. (2015). Automatic translation of syntagms into Signagrams for risk prevention. *Management des technologies organisationnelles*, 5, 23–37.

Kraut, R.E. and Resnick, P. (2012). *Building Successful Online Communities: Evidence-based Social Design*. MIT Press, Massachusetts.

Löwstedt, A. (2018). Communication ethics and globalization. *Communication and Media Ethics*, 26, 367–390.

Ma, C.M., Zhuang, Y., Fong, S. (2011). Information sharing over collaborative social networks using xacml. *IEEE 8th International Conference on e-Business Engineering*, 161–167.

McShane, M., Nirenburg, S., Beale, S. (2005). An NLP lexicon as a largely language-independent resource. *Machine Translation*, 19(2), 139–173.

Miller, G.A. (1998). *WordNet: An Electronic Lexical Database*. MIT Press, Massachusetts.

Morin, D.B., Mierau, D.R., Van Horn, M., Dofter, D.S., Lewandowski, M., Trinh, D., Jackson, M., Kirsh, L., Matteson, M.M., Crosby, J., Path Inc (2012). Method and system for a personal network. U.S. Patent Application 12/945,743.

National Commission on Informatics and Liberty (2020). NCIL [Online]. Available at: https://www.cnil.fr/en/home.

Neurath, M. (1974). Isotype. *Instructional Science*, 3(2), 127–150.

Noy, N.F. and McGuinness, D.L. (2001). Ontology development 101: A guide to creating your first ontology [Online]. Available at: http://www.corais.org/sites/default/files/ontology_development_101_aguide_to_creating_your_first_ontology.pdf [Accessed 9 April 2020].

Palmirani, M., Martoni, M., Rossi, A., Bartolini, C., Robaldo, L. (2018a). PrOnto: Privacy ontology for legal reasoning. In *Electronic Government and the Information Systems Perspective*, Kő, A. and Francesconi, E. (eds). Springer, Cham.

Palmirani, M., Martoni, M., Rossi, A., Bartolini, C., Robaldo, L. (2018b). Legal ontology for modelling GDPR concepts and norms. *JURIX*, 91–100.

Proulx, S. (2012). L'irruption des médias sociaux : enjeux éthiques et politiques. In *Médias sociaux : enjeux pour la communication*, Proulx, S., Millette, M., Heaton, L. (eds). Presses de l'Université du Québec, Quebec.

Rainie, L. and Wellman, B. (2012). *Networked: The New Social Operating System*. MIT Press, Massachussetts.

Rossi, A. and Palmirani, M. (2019). DaPIS: An ontology-based data protection icon set. *Knowledge of the Law in the Big Data Age*, 317, 181–195.

Seme, Y. and Microsoft Corp (2003). Method and system for translating instant messages. U.S. Patent Application 10/035,085.

Spiliotopoulos, D., Tzoannos, E., Cabulea, C., Frey, D. (2013). Digital archives: Semantic search and retrieval. In *Human-Computer Interaction and Knowledge Discovery in Complex, Unstructured, Big Data*, Holzinger, A. and Pasi, G. (eds). Springer, Berlin, Heidelberg.

Suojanen, T., Koskinen, K., Tuominen, T. (2014). *User-Centered Translation*. Taylor & Francis Group, Routledge, London.

Takasaki, T. (2006). PictNet: Semantic infrastructure for pictogram communication. *The Third International WordNet Conference (GWC-06)*, 279–284.

Takasaki, T. and Mori, Y. (2007). Design and development of a pictogram communication system for children around the world. In *International Workshop on Intercultural Collaboration*, Ishida, T., Fussell, S.R., Vossen, P.T.J.M. (eds). Springer, Berlin, Heidelberg.

Turkle, S. (2011). *Life on the Screen: Identity in the Age of the Internet*. Simon and Schuster, New York.

United Nations Economic Commission for Europe – Transport Division (2020). Road traffic and road signs and signals agreements and conventions [Online]. Available at: https://www.unece.org/fileadmin/DAM/trans/conventn/Conv_road_signs_2006v_EN.pdf [Accessed 9 April 2020].

Waheed, H., Anjum, M., Rehman, M., Khawaja, A. (2017). Investigation of user behavior on social networking sites. *PloS One*, 12(2). DOI: 10.1371/journal.pone.0169693.

Wellman, B. (2002). Little boxes, glocalization, and networked individualism. In *Digital Cities II: Computational and Sociological Approaches*, Tanabe, M., Besselaar, P., Ishida, T. (eds). Springer, Berlin, Heidelberg.

Wellman, B. and Gulia, M. (1999). Net-surfers don't ride alone: Virtual communities as communities. *Networks in the Global Village: Life in Contemporary Communities*, 10(3), 34–60.

Yang, C.Y. and Lin, H.Y. (2010). An instant messaging with automatic language translation. *3rd IEEE International Conference on Ubi-Media Computing*, 312–316.

6

A New Approach of Texts and Writing Normalization for Arabic Knowledge Organization

6.1. Introduction

The normalization of the Arabic language and its writing are a necessity for its automatic processing. This mainly concerns the linguistic composition elements: syntactic, semantic, stylistic and orthographic structures (Fadili 2020a, 2020b). In order to contribute in this context, we propose an unsupervised approach based on deep learning implementing a normalization support system of writing in general, and in the context of written Arabic texts.

The approach is based mainly on the improvement of the Bi-RNN (bidirectional recurrent neural network) model and its contextual implementation Bi-LSTM (bidirectional long short-term memory) for sequence prediction. We have implemented mechanisms to capture the writing signals encoded through the different layers of the network for optimizing the prediction of the next structures and the next words completing the sentences in writing progress or transforming texts already written into their normalized forms. The latent structures and the reference spelling are those learned from the training corpus considering their importance and relevance: integration of the notions of attention and point of view in Bi-LSTM.

Chapter written by Hammou FADILI.

This chapter is structured as follows: section 6.3 presents the learning model, section 6.4 defines the methodological and technological elements implementing the approach, section 6.5 is devoted to the generation of the dataset for learning and, finally, section 6.6 presents the tests and the results obtained.

6.2. Motivation

The language writing normalization that we are talking about concerns all aspects related to syntactic, semantic, stylistic and orthographic structures (Fadili 2020a, 2020b). This is a complex task, especially in the case of languages used on the Internet, such as Arabic. Proposing solutions that can help authors and new users of such languages to respect and popularize their standards could be an important element in the process of standardizing their writings.

The rapid return and development of artificial intelligence makes such solutions possible: unsupervised approaches that do not require prerequisites or preprocessed data can be used to code the latent science contained in the studied corpora as the basis of machine learning and as references for normalization.

These are the elements that motivated us to study and propose an unsupervised approach, allowing us to detect and use the "NORMALIZATION" coded and contained in "well-written" texts in terms of spelling, composition and structure language.

Our contribution consists, on the one hand, of adapting and instantiating the contextual data model (see Fadili 2017) and, on the other hand, of making improvements to the basic Bi-LSTM in order to circumvent their limits in the management of "contextual metadata".

6.3. Using a machine learning model

Several studies have shown that deep learning has been successfully exploited in many fields, including that of automatic natural language processing (NLP). One of the best implementations is the generation of dense semantic vector spaces (Mikolov 2013).

Other networks such as RNN (recurrent neural networks) have also been improved and adapted to support the recurrent and sequential nature of NLP: each state is calculated from its previous state and new entry. These networks have the advantage of propagating information in both directions: towards the input and output layers, thus reflecting an implementation of neural networks, close to the functioning of the human brain where information can be propagated in all directions by exploiting the memory principle (see the LSTM version of RNN in the following), via recurrent connections propagating the information of ulterior learning (the memorized information).

These are the characteristics that allow them to take better care of several important aspects of natural language. Indeed, they have this ability to capture latent syntactic, semantic, stylistic and orthographic structures, from the order of words and their characteristics, unlike other technologies such as those based on the concept of bag of words (BOW) where the order is not considered, obviously involving loss of the associated information.

In serial RNNs, each new internal and output state simply depends on the new entry and the old state.

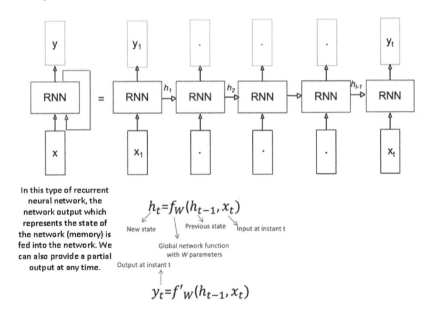

In this type of recurrent neural network, the network output which represents the state of the network (memory) is fed into the network. We can also provide a partial output at any time.

$$h_t = f_W(h_{t-1}, x_t)$$

New state   Previous state   Input at instant t

Global network function with W parameters

Output at instant t

$$y_t = f'_W(h_{t-1}, x_t)$$

Figure 6.1. *How an RNN works. For a color version of this figure, see www.iste.co.uk/sidhom/systems.zip*

RNNs can also be stacked and bidirectional, and the simple equations shown in Figure 6.1 can be redefined for the two learning directions according to the model presented below.

$Forward\ propagation: h_t^l$

$$= \tanh W_t^l \begin{pmatrix} h_t^{l-1} \\ h_{t-1}^l \end{pmatrix}, h_t^l$$

$\in \mathbb{R}^n, W^l\ matrix\ of\ [n \times 2n]$ dimensions

$Backward\ propagation: h'^l_t$

$$= \tanh W'^l_t \begin{pmatrix} h_t^{l-1} \\ h'^l_{t+1} \end{pmatrix}, h_t^l, h'^l_t$$

$\in \mathbb{R}^n, W'^l\ matrix\ of\ [n \times 2n]\ dimensions$

$Output: y_t = \tanh W_t \begin{pmatrix} h_t^l \\ h'^l_t \end{pmatrix}, y_t$

$\in \mathbb{R}^n, W^l\ matrix\ of\ [n \times 2n]\ dimensions$

The training of multi-layer RNNs is done, as for the other types of networks, by minimizing the error (difference between the desired output and the output obtained), which can be obtained by the back-propagation of the error and the descent of the gradient. It can be demonstrated mathematically that the depth of RNNs, which can be high, because of sequential nature of paragraphs, texts, documents, etc., generally depends on the number of words to be processed at a time; this can provoke:

– either the Vanishment of Gradient in the first layers and the end of learning from a certain depth. When we must multiply the gradient, a maximum number of times, by a weight w/|w|<1;

– or the Explosion of Gradient always in the first layers and the end of learning from a certain depth. When we must multiply the gradient, a maximum number of times, by a weight w/|w|>1.

The LSTM architecture makes it possible to remedy these problems (Hochreiter and Schmidhuber 1997). It is based on finer control of the information flow in the network, thanks to three gates: the forget gate, which decides what to delete from the state (ht-1, xt); the input gate, which chooses what to add to the state; and the output gate, which chooses what we must keep from the state (see equations shown in Figure 6.2).

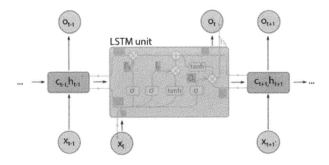

Figure 6.2. *How a basic LSTM works. For a color version of this figure, see www.iste.co.uk/sidhom/systems.zip*

$$F_t = \sigma(W_F x_t + U_F h_{t-1} + b_F) \qquad \text{(forget gate)}$$
$$I_t = \sigma(W_I x_t + U_I h_{t-1} + b_I) \qquad \text{(input gate)}$$
$$O_t = \sigma(W_O x_t + U_O h_{t-1} + b_O) \qquad \text{(output gate)}$$
$$c_t = F_t \circ c_{t-1} + I_t \circ \tanh(W_c x_t + U_c h_{t-1} + b_c)$$
$$h_t = O_t \circ \tanh(c_t)$$
$$o_t = f(W_o h_t + b_o)$$

These equations that define the learning process of an LSTM express the fact that this kind of network makes it possible to cancel certain useless information and to reinforce others having a great impact on the results. We can also show by mathematical calculations that this architecture, in addition to the optimization of the calculations in the network, makes it possible to solve the problems linked to the vanishing and explosion of the gradient. This is what motivated our choice to use and improve this model by adapting it to our needs. We also integrated the notion of perspective or point of view of analysis as well as the notion of attention in the general process.

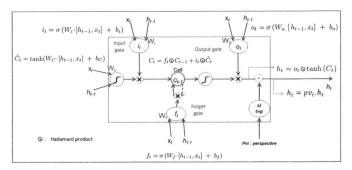

Figure 6.3. *Architecture of an improved LSTM. For a color version of this figure, see www.iste.co.uk/sidhom/systems.zip*

Therefore, our model allows us to control the flow in the "Context" i.e.:

– what to forget from the state;

– what to use from the state;

– what to send to the next state;

– according to a point of view or perspective;

– by paying attention to relevant information.

The latter notion is represented outside the internal architecture of the LSTM model.

6.4. Technological elements integration

Our approach has two main objectives:

– help users/authors to normalize their writing during the writing process;

– normalize already written texts as an early preprocessing step in an automatic analysis.

In the first case, our approach aims to predict and suggest supplementing the sentences being written with the most relevant words in their standardized orthographic forms and following a very specific linguistic structure (learned from texts of the training corpus). In other words, to a sequence of words that the user is typing, the system proposes a sequence of words, sentences or parts of standardized sentences.

In the second case, our approach aims to normalize the already written texts as an early pretreatment step in a semantic automatic analysis process. This is performed by transforming the texts, sentences and words into their normalized forms according to the orthographic, linguistic, stylistic structures, etc. learned from the training corpus.

In both cases, this requires the use of the sequence-to-sequence or seq2seq version of Bi-LSTMs.

The proposed architecture consists of the following main layers:

– encoding:

  - pretreatment,

- internal representation,

- domain (point of view);

– decoding:

- new internal representation (calculated),

- attention,

- prediction.

A perspective is a set of words characterizing an analysis point of view (multidimensional space).

Let p be the number of words in the current sentence

Let i be the current perspective index

Let P be a perspective defined by the dimensions $p_1, p_2, ..., p_l$

Let H be the matrix composed by the hidden representations h,h_2,...,h_p generated by the LSTM

Consider $pv_i = \sum_1^p d_j$ where $d_j = \sum_1^l d_k$ and $d_k = \frac{p_k.h_j}{\|p_k\|\|h_j\|}$

– pv_i is a first weight in the prediction calculation, based on the sum of the cosine distances between the hidden representations generated by the LSTM and the dimensions of the targeted analysis perspective.

This makes it possible to obtain a first transformation, weighted by the comparison to the view considered, of the internal representations of the meanings of the sentence:

$H' = \left(h'_1, h'_2, ..., h'_p\right), where\ h'_j = d_j.h_j$

$d_j\ is\ the\ sum\ of\ the\ cosine\ ditances\ of\ hi\ at\ each\ dimension.$

$\propto_d = softmax\left(\sum_1^p at_i\right)$

at_i

$= w_i^T.h'_i\ constitutes\ the\ attention\ learned\ by\ (w_i)\ for\ each\ h'_i$

$prediction = H \propto_d^T\ represents\ the\ output\ of\ the\ system.$

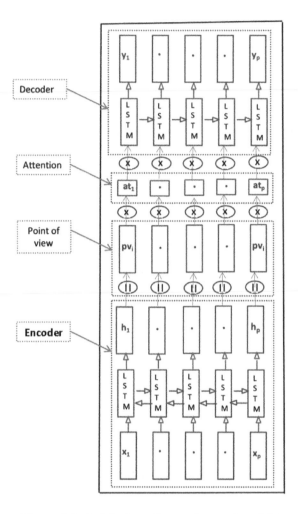

Figure 6.4. *Architecture. For a color version of this figure, see www.iste.co.uk/sidhom/systems.zip*

6.5. Corpus and dataset

The corpus was built up by the collection of many documents in Arabic, obtained mainly from "well-written" institutional websites. The learning data model was obtained from a simplified version of the language model and the extended semantic model (Fadili 2017) and projections of the initial vector representations of the words, relating to a space of large dimensions (vocabulary size), in a reduced dimension semantic space using the

Word2vec technology (w2v). The goal is to create an enriched model of instances adapted to the context of language normalization, with a reasonable size vector representation, essential for optimizing calculations and processing. The instantiation of the model was done by splitting the texts into sentences, and the generation of the enriched n-grams context windows for each word:

- 2-grams;

- 3-grams;

- 4-grams;

- 5-grams.

This first version of the instances is enriched with other parameters to support the syntactic structures and the thematic distributions. We associated each word in the dataset with its grammatical category (Part Of Speech – POS) as well as its thematic distribution (Topics) obtained by LDA (Latent Dirichlet Allocation) (Bei et al. 2003).

The other latent structures are integrated by the coding of the sequential nature of the texts by the LSTM. Similarly, the spelling is coded from the training corpus. It is this dataset that is provided to our augmented and improved Bi-LSTM model.

This model has the advantage of considering, in fine, the local context (n-grams) and the global context (themes), long-term and word order memories, which encode the latent structures of the texts (syntactic, semantic, etc.).

6.6. Experiences and evaluations

We have developed several modules and programs, implementing the elements of the general process, including mainly:

- automatic extraction of the various characteristics (Features);

- implementation of a deep learning system based on Bi-LSTM endowed with attention and domain mechanisms.

We have also designed an environment centralizing access to all modules:

– integration of all the elements in a single workflow implementing all the processing modules.

Feature extraction

We have developed and implemented three modules that run in the same "Python-Canvas" environment and exploited a process based on a "General MetaJoint" of the same environment to make the link between different components.

Without going into detail, the three modules consist, in order, of extracting the following characteristics:

– The first module allows us to extract the "linguistic context" of each word through a sliding window of size n.

– The second module allows us to grammatically annotate all the words in the text: perform POS Tagging (Part Of Speech Tagging).

– The third implements the "Topic modeling and LDA" technology in order to automatically extract the studied domain(s).

Workflow

The implementation of the workflow was done in the Orange-Canvas platform. It automates the chaining of results from previous modules for the extraction of characteristics and the instances ready to be used for learning.

Improved Bi-LSTM neural network

We developed from scratch a personalized Bi-LSTM neural network and adapted it to our needs. It is a neural network (multi-layer and recurrent Bi-LSTM) endowed with attention and domain mechanisms.

In order to ensure the honesty of the system, we separated the generated learning data into three parts:

– The first part was for validation on 20% of the dataset, in order to optimize the hyper-parameters of the system: the learning step, the type of the activation function and the number of layers.

– The rest of the dataset was divided into two parts:

- 60% for training, in order to estimate the best coefficients (wi) of the function of the neural network, minimizing the error between the actual and desired outputs;

- 20% for tests, in order to assess the performance of the system.

During all the learning phases, the system is autonomous, which can generate the features of the text (Features) for training and perform the learning process (validation, training and tests).

6.6.1. *Results*

=== Evaluation on test split ===

Time taken to test the model on test split: 0.75 seconds

=== Summary ===

Correctly classified instances	*455*	*83.7937%*
Incorrectly classified instances	*88*	*16.2063%*
Kappa statistic	*0.8323*	
Mean absolute error	*0.0021*	
Root mean squared error	*0.0302*	
Relative absolute error	*21.9222%*	
Root relative squared error	*44.2078%*	
Total number of instances	*543*	

After training the system on the constituted corpus, we conducted tests on a small text of 543 words. The system succeeded in correctly predicting the location and the punctuation of 455 words in the different sentences of the text, considering the latent structures coded from the training texts. The system, however, failed on the remaining 88 words and punctuation.

For more information on these results, refer to WEKA (Hall et al. 2009).

6.7. Conclusion

Our approach is an important contribution to the field of Arabic NLP, on the one hand, and to the organization of textual knowledge of the same language, on the other.

The knowledge generated today on the Web is poorly written, organized, structured, etc., since web users have the freedom to express themselves without any orthographic, stylistic and syntactic constraints. Hence, we need sophisticated and innovative tools, especially in the case of Arabic, to conduct its treatments. In addition, the organization of such knowledge is an even more complex problem, especially in the context of Big Data where processing by a human is almost impossible to achieve.

Our approach, which was designed to respond to all these problems, enabled us to obtain good results, as evidenced by the various statistics provided in section 6.6 (Fadili 2020a, 2020b).

As described above, we adjusted several hyper-parameters to optimize the system and generate all the parameters of the instances. On the entire corpus obtained, we validated the approach on 20% of the corpus, trained the system on approximately 60% and tested on approximately 20%. The various measurements show the good performance of the approach in terms of accuracy and loss.

This is mainly due to improvements made to the integration of certain aspects of the language and to extensions implemented for Bi-LSTM:

– language model and its instantiation;

– introduction of the concepts of attention and domain.

6.8. References

Agarwal, A., Biadsy, F., Mckeown, K.R. (2009). Contextual phrase-level polarity analysis using lexical affect scoring and syntactic n-grams. In *Proceedings of the 12th Conference of the European Chapter of the Association for Computational Linguistics*, Lascarides, A., Gardent, C., Nivre, J. (eds). Association for Computational Linguistics, Athens.

Amiri, H. and Chua, T.S. (2012). Sentiment classification using the meaning of words. *Twenty-Sixth AAAI Conference on Artificial Intelligence*, July 22–26.

Augenstein, I. (2012). LODifier: Generating linked data from unstructured text. In *The Semantic Web: Research and Applications*, Simperl, E., Cimiano, P., Polleres, A., Corcho, O., Presutti, V. (eds). Springer, Berlin, Heidelberg.

Besag, J. (1986). On the statistical analysis of dirty pictures. *Journal of the Royal Statistical Society. Series B (Methodological)*, 48(3), 259–302.

Boudia, M.A., Hamou, R.M., Amine, A. (2016). A new approach based on the detection of opinion by Sentiwordnet for automatic text summaries by extraction. *International Journal of Information Retrieval Research (IJIRR)*, 6(3), 19–36.

Boury-Brisset, A.-C. (2013). Managing semantic big data for intelligence. In *Stids*, Laskey, K.B., Emmons, I., da Costa, P.C.G. (eds) [Online]. Available at: CEUR-WS.org.

Cambria, E., Schuller, B., Xia, Y., Havasi, C. (2013). New avenues in opinion mining and sentiment analysis. *IEEE Intelligent Systems*, 28(2), 15–21.

Chan, J.O. (2014). An architecture for big data analytics. *Communications of the IIMA*, 13.2(2013), 1–13.

Christian, B., Tom, H., Kingsley, I., Tim, B.L. (2008). Linked data on the web (LDOW2008). *Proceedings of the 17th International Conference on World Wide Web*, 1265–1266.

Curran, J.R., Clark, S., Bos, J. (2007). Linguistically motivated large-scale NLP with C&C and boxer. *Proceedings of the ACL 2007 Demonstrations Session (ACL-07 Demo)*, Prague.

Das, T.K. and Kumar, P.M. (2013). Big data analytics: A framework for unstructured data analysis. *International Journal of Engineering and Technology*, 5(1), 153–156.

DBpedia (2022). Global and Unified Access to Knowledge Graphs [Online]. Available at: http://wiki.dbpedia.org/.

Denecke, K. (2008). Using sentiwordnet for multilingual sentiment analysis. *IEEE 24th International Conference on Data Engineering Workshop, ICDEW*, 7–12 April.

Dimitrov, M. (2013). From big data to smart data. *Semantic Days* [Online]. Available at: https://pt.slideshare.net/marin_dimitrov/from-big-data-to-smart-data-22179113.

Duan, W., Cao, Q., Yu, Y., Levy, S. (2013). Mining online user-generated content: Using sentiment analysis technique to study hotel service quality. *46th Hawaii International Conference on System Sciences*, 7–10 January.

Esuli, A. and Sebastiani, F. (2007). SentiWordNet: A high-coverage lexical resource for opinion mining. *Evaluation. 5th Conference on Language Resources and Evaluation (LREC'06)*, January.

Fadili, H. (2013). Towards a new approach of an automatic and contextual detection of meaning in text: Based on lexico-semantic relations and the concept of the context. *IEEE-AICCSA*, 27–30 May.

Fadili, H. (2016a). Towards an automatic analyze and standardization of unstructured data in the context of big and linked data. *MEDES*, 223–230.

Fadili, H. (2016b). Le machine Learning : numérique non supervisé et symbolique peu supervisé, une chance pour l'analyse sémantique automatique des langues peu dotées. *TICAM*, Rabat.

Fadili, H. (2017). Use of deep learning in the context of poorly endowed languages. *24e Conférence sur le Traitement Automatique de la Langue Naturelle (TALN)*, Orléans.

Fadili, H. (2020a). Semantic mining approach based on learning of an enhanced semantic model for textual business intelligence in the context of big data. *OCTA Multi-conference Proceedings: Information Systems and Economic Intelligence (SIIE)* 208 [Online]. Available at: https://multiconference-octa.loria.fr/multiconference-program/.

Fadili, H. (2020b). Deep learning of latent textual structures for the normalization of Arabic writing. *OCTA Multi-conference Proceedings: International Society for Knowledge Organization (ISKO-Maghreb)*, 170 [Online]. Available at: https://multiconference-octa.loria.fr/multiconference-program/.

Fan, J., Han, F., Liu, H. (2014). Challenges of big data analysis. *National Science Review*, 1(2), 293–314.

Frijda, N.H., Mesquita, B., Sonnemans, J., Van Goozen, S. (1991). The duration of affective phenomena or emotions, sentiments and passions. In *International Review of Studies on Emotion*, Strongman, K.T. (ed.). Wiley, New York.

Gupta, A., Viswanathan, K., Joshi, A., Finin, T., Kumaraguru, P. (2011). Integrating linked open data with unstructured text for intelligence gathering tasks. *Proceedings of the Eighth International Workshop on Information Integration on the Web*, March.

Hall, M., Frank, E., Holmes, G., Pfahringer, B., Reutemann, P., Witten, I.H. (2009). The WEKA data mining software: An update. *SIGKDD Explorations*, 11(1).

Hiemstra, P.H., Pebesma, E.J., Twenhofel, C.J.W., Heuvelink, G.B.M. (2009). Real-time automatic interpolation of ambient gamma dose rates from the Dutch radioactivity monitoring network. *Computers & Geosciences*, 35(8), 1711–1721.

Hochreiter, S. and Schmidhuber, J. (1997). LSTM can solve hard long time lag problems. In *Advances in Neural Information Processing Systems 9*, Mozer, M.C., Jordan, M.I., Petsche, T. (eds). MIT Press, Cambridge, MA.

Hopkins, B. (2016). Think you want to be "data-driven"? Insight is the new data [Online]. Available at: https://go.forrester.com/blogs/16-03-09-think_you_want_to_be_data_driven_insight_is_the_new_data/.

Hu, M. and Liu, B. (2004). Mining and summarizing customer reviews. *Proceedings of the Tenth ACM SIGKDD International Conference on Knowledge Discovery and Data Mining*, 168–177.

Hung, C. and Lin, H.K. (2013). Using objective words in SentiWordNet to improve sentiment classification for word of mouth. *IEEE Intelligent Systems*, 1.

Kamp, H. (1981). A theory of truth and semantic representation. In *Formal Semantics – the Essential Readings*, Portner, P. and Partee, B.H. (eds). Blackwell, Oxford.

Khalili, A., Auer, S., Ngonga, Ngomo, A.-C. (2014). conTEXT – Lightweight Text Analytics using Linked Data. *Extended Semantic Web Conference (ESWC 2014)*, 628–643.

Khan, E. (2013). Addressing big data problems using semantics and natural language understanding. *12th Wseas International Conference on Telecommunications and Informatics (Tele-Info '13)*, Baltimore, 17–19 September.

Khan, E. (2014). Processing big data with natural semantics and natural language understanding using brain-like approach. *International Journal of Computers and Communication*, 8.

Le, Q. and Mikolov, T. (2014). Distributed representations of sentences and documents. *Proceedings of the 31st International Conference on Machine Learning (ICML-14)*, 32(2), 1188–1196.

Mikolov, T. (2013). Statistical language models based on neural networks. PhD Thesis, Brno University of Technology.

Miller, G.A. (1995). WordNet: A lexical database for English. *Communications of the ACM*, 38(11), 39–41.

Minelli, M., Chambers, M., Dhiraj, A. (2013). *Big Data, Big Analytics: Emerging Business Intelligence and Analytic Trends for Today's Businesses*. Wiley, New York.

Nassirtoussi, A.K., Aghabozorgi, S., Wah, T.Y., Ngo, D.C.L. (2014). Text mining for market prediction: A systematic review. *Expert Systems with Applications*, 41(16), 7653–7670.

Neelakantan, A., Shankar, J., Passos, A., McCallum, A. (2014). Efficient non-parametric estimation of multiple embeddings per word in vector space. *Conference on Empirical Methods in Natural Language Processing*, October.

Paltoglou, G. and Thelwall, M. (2012). Twitter, MySpace, Digg: Unsupervised sentiment analysis in social media. *ACM Transactions on Intelligent Systems and Technology (TIST)*, 3(4), 66.

Park, C. and Lee, T.M. (2009). Information direction, website reputation and eWOM effect: A moderating role of product type. *Journal of Business Research*, 62(1), 61–67.

Rao, D. and Ravichandran, D. (2009). Semi-supervised polarity lexicon induction. *Proceedings of the 12th Conference of the European Chapter of the Association for Computational Linguistics*, 675–682.

Rusu, D., Fortuna, B., Mladenić, D. (2011). Automatically annotating text with linked open data. *4th Linked Data on the Web Workshop (LDOW 2011), 20th World Wide Web Conference*.

Shand, A.F. (2014). *The Foundations of Character: Being a Study of the Tendencies of the Emotions and Sentiments*. HardPress, Miami.

Singh, V.K., Piryani, R., Uddin, A., Waila, P. (2013). Sentiment analysis of textual reviews. Evaluating machine learning, unsupervised and SentiWordNet approaches. *5th IEEE International Conference on Knowledge and Smart Technology (KST)*, January 31–February 1.

Taboada, M., Brooke, J., Tofiloski, M., Voll, K., Stede, M. (2011). Lexicon-based methods for sentiment analysis. *Computational Linguistics*, 37(2), 267–307.

Tao, C., Song, D., Sharma, D., Chute, C.G. (2003). Semantator: Semantic annotator for converting biomedical text to linked data. *Journal of Biomedical Informatics*, 46(5), 882–893. DOI: 10.1016/j.jbi.2013.07.003.

Tumasjan, A., Sprenger, T.O., Sandner, P.G., Welpe, I.M. (2010). Predicting elections with twitter: What 140 characters reveal about political sentiment. *ICWSM*, 10(1), 178–185.

Wang, H., Can, D., Kazemzadeh, A., Bar, F., Narayanan, S. (2012). A system for real-time Twitter sentiment analysis of 2012 US presidential election cycle. *Proceedings of the ACL 2012 System Demonstrations*, Jeju Island, July.

Zhang, W. and Skiena, S. (2009). Improving movie gross prediction through news analysis. *Proceedings of the 2009 IEEE/WIC/ACM International Joint Conference on Web Intelligence and Intelligent Agent Technology-Volume 01*, 15–18 September.

Zhao, J., Dong, L., Wu, J., Xu, K. (2012). Moodlens: An emoticon-based sentiment analysis system for Chinese tweets. *Proceedings of the 18th ACM SIGKDD International Conference on Knowledge Discovery and Data Mining*, August.

Zhou, X., Tao, X., Yong, J., Yang, Z. (2013). Sentiment analysis on tweets for social events. *17th IEEE International Conference on Computer Supported Cooperative Work in Design (CSCWD)*, June.

Ebola Epidemic in the Congo 2018–2019: How Does Twitter Permit the Monitoring of Rumors?

7.1. Introduction

Since the year 2000, the research in human and social sciences has been revolutionized by the advent of the digital age. Generally, humanities are undergoing a deep change due to this digital revolution (Quentin and Citton 2015).

From these cultural and social paradigms, a new field emerges, at the crossing of computing, human and social sciences: digital humanities. This field particularly investigates the social and cultural impact of information and communication technologies and plays an active role in the design, implementation, questioning and subversive nature of these technologies today (Rieffel 2014).

Current studies are focused on studying the subversive defaults of these technologies, in particular those of societal culture, humanity and civilizations (Breton 2000).

Few papers analyze positive applications of such technologies for humanities and individuals, particularly to solve social and societal issues. In this context, health is a major social issue which is deeply revolutionized by

Chapter written by Marc TANTI.

these technologies. Health studies tend to report potential abuse of digital technology, especially in terms of ethics (Colloc 2014).

However, digital technology has had a positive impact in the field of health, specifically concerning the management of complex sanitary problems, for example, epidemics. But a few papers have analyzed this positive impact, particularly the positive repercussions of social networks in the health domain.

The last major Ebola epidemic in Africa took place in 2014–2015. This epidemic mainly affected three West African countries (Guinea, Sierra Leone, Liberia). It killed 20,000 people. A number of articles have investigated the rumors that circulated on Twitter during this epidemic. For example, the article by Fung et al. (2016) pointed out that these media disseminated false information about the treatment of the disease, such as bathing in salt water to cure it.

The article by Jin et al. (2014) also pointed out that these media were behind fake news of a snake at the origin of the epidemic. This article was also listed in the top 10 rumors about the Ebola epidemic circulating on Twitter (see Figure 7.1).

Table 1. Top 10 Ebola-related rumors by Tweet volume from 28 September to 18 October 2014.		
Rumor no.	**Content**	**Label**
1	Ebola vaccine only works on white people	White
2	Ebola patients have risen from the dead	Zombie
3	Ebola could be airborne in some cases	Airborne
4	Health officials might inject Ebola patients with lethal substances	Inject
5	There will be no 2016 election and complete anarchy	Vote
6	The US government owns a patent on the Ebola virus	Patent
7	Terrorists will purposely contract Ebola and spread it around	Terrorist
8	The new iPhone 6 is infecting people with Ebola	iPhone
9	There is a suspected Ebola case in Kansas City	Kansas
10	Ebola has been detected in hair extensions	Hair

Figure 7.1. *Top 10 rumors circulating on Twitter (Jin et al. 2014)*

Ebola Virus Disease is still raging today in the Democratic Republic of Congo (DRC), since August 1, 2018. It has killed more than 2,050 people. It is the second largest epidemic after that of West Africa (Ebola Outbreak Epidemiology Team 2018; Medley et al. 2020).

In addition, numerous studies have shown that Twitter is used by public health organizations, in particular to inform, educate or monitor the state of health of populations, particularly in the event of a disaster (Hart et al.

2017). However, no studies have been conducted to determine who is communicating about the current Ebola epidemic in the DRC and what types of tweets/rumors are being circulated (Tanti et al. 2012).

To answer this question, we conducted an analysis on Twitter via the Radarly® software, over a period from January 4, 2019 to July 7, 2019. The keywords Ebola and #Ebola were used, with a filter on the French language. A total of 17,282 tweets were collected and classified via the software. The tool also extracted and represented the knowledge in a cartographic way (volume of publications/time, etc.). It also made it possible to identify the dominant themes in the form of clusters. The tone of the messages has also been determined.

After a description of the methodology used, this chapter presents the main results obtained, in particular the fact that several actors communicate around the epidemic, in particular the general public, experts, politicians and the press.

7.2. Materials and methods

To carry out this study, we used the Radarly® social media monitoring software marketed by Linkfluence (see: https://radarly.linkfluence.com/ login) and operating in SAAS mode (see Figure 7.2). This software is accessible online on subscription and allows us to collect data on the social web (Tweet, Facebook, Instagram).

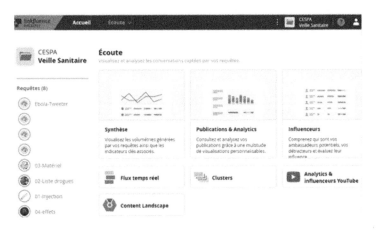

Figure 7.2. *Radarly interface. For a color version of this figure, see www.iste.co.uk/sidhom/systems.zip*

The software also makes it possible to represent the results in a cartographic manner, in particular in the form of clusters of dominant subjects. It also makes it possible to carry out analyses of the tone of the messages published. It identifies "influencers" (people or groups who speak on a given topic or theme). It allows the export of data in .csv format to deduce statistics (see Figure 7.3).

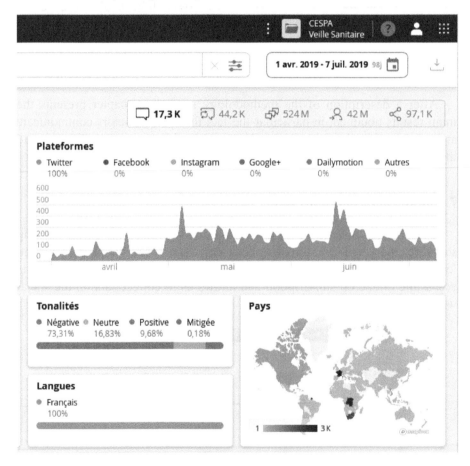

Figure 7.3. *Graphical representations of Radarly® data. For a color version of this figure, see www.iste.co.uk/sidhom/systems.zip*

To collect data from Radarly® software, we applied the monitoring process and the methodology (see Figure 7.4) developed by Tanti et al. (2012) in the article entitled *"Pandémie grippale 2009 dans les armées : l'expérience du veilleur"*, which includes six stages: definition of monitoring themes, identification, collection, analysis, synthesis and distribution of documents.

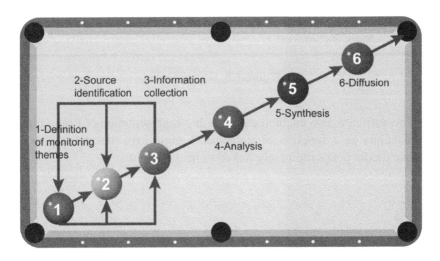

Figure 7.4. *Monitoring process (Tanti et al. 2012). For a color version of this figure, see www.iste.co.uk/sidhom/systems.zip*

Concerning the first step, the definition of the themes of monitoring, the keywords Ebola and #Ebola were used in the request, with a filter on the French language.

Concerning the second and third stages, the identification and selection of documentary sources, we only selected the collection of tweets on the social network Twitter via the platform.

The analysis step was done using Radarly® cartographic analysis and representation functionalities.

We analyzed tweets posted on Twitter over a period from April 1, 2019 to July 7, 2019. A total of 17,282 tweets were collected, classified and categorized via the software.

Figure 7.5 summarizes the number of data collected and analyzed.

> ⁎ Number of posts: 17,282 posts
> ⁎ Number of posts and retweets: 44,166 posts and retweets
> ⁎ Estimated reach: 42 million estimated posts. This is the estimated number of Internet users having seen posts or retweets
> ⁎ Amount of engagement: 97,122: this is the total number of engagement actions for each post (likes, retweets, comments, shares, favorites etc.)

Figure 7.5. *Data collected*

The software also enabled cartographic representations of the volume of publications as a function of time (see Figure 7.6), making it possible to deduce media peaks intimately linked to health events.

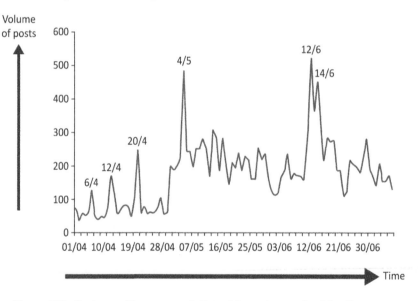

Figure 7.6. *Cartographic representation of the volume of publications as a function of time (January 4, 2019 to July 7, 2019). For a color version of this figure, see www.iste.co.uk/sidhom/systems.zip*

7.3. Results

The requests have identified the actors who communicate on Twitter concerning the Ebola disease, which has been raging since January 8, 2018 in the DRC and has left more than 2,000 people dead.

The filter on the French language made it possible to only select tweets written in French. The analysis also made it possible to identify the messages conveyed.

The main players found were:

– the general public, mainly Congolese citizens and associations;

– experts and health organizations (Ministry of Health of the Congo, WHO, etc.);

– the press, mainly Congolese, etc.;

– politicians, mainly Congolese.

7.3.1. *Regarding the general public, the citizens*

With the advent of social networks, the forms and possibilities of communication offered to citizens have been considerably multiplied. Citizens and the general public can easily and instantly put their personal opinions online on their own site, on Facebook or other networks. A study even sees Twitter as a "public opinion barometer" (Boyadjian 2014). This advent of the social web has ushered in a new era, the "e-democracy". This sociological concept, developed in the early 2000s, underpins three impacts on citizen life: rapid access to a mass of information, communication and citizen participation favored by new technologies and more direct deliberative action (Rodota 1999).

In our study, it is mainly Congolese citizens and associations who speak out on the epidemic on the social network. In the Congolese population, there are divided opinions and two populations: a population that "believes" and a population that does not "believe" in the disease.

The party that "believes" accepts the disease and the epidemic and considers it as a public health problem. It adheres to medical treatment and preventive measures. It relays messages of scientific information, education and awareness, etc. As an example, we can cite a tweet relaying the effectiveness of the vaccine (see Figure 7.7).

POLITICO.CD ✔
@politicocd
96.9 k followers

20 avril 2019 - 11:27

#RDC/#Ebola: Les traitements expérimentaux et les vaccins ont réduit le taux de mortalité
https://t.co/hKLqXlLihB https://t.co/hKLqXlLihB

Figure 7.7. *Tweet relaying the effectiveness of the vaccine. For a color version of this figure, see www.iste.co.uk/sidhom/systems.zip*

The rest of the population is less "gullible" and denies the disease. It constitutes the majority of the tweets found (high negative tone). Thus, despite the efforts of response teams since the start of the epidemic in August 2018, this population considers the disease as a plot to destabilize the country. This population spreads rumors. They accuse, for example, the laboratories or the WHO of having created the virus (see Figure 7.8).

Dimandja mukumadi
@Dimandjamukuma2
5 followers

14 mai 2019 - 08:58

Epidemie d'ebola sous disent que c par le main sale ?mensonge ce dans le laboratoire imperialiste est transmis par un organisation de sante "OMS" pour le propager ds le cour d'eau en afrique precisement au congo

Figure 7.8. *Tweet accusing laboratories of creating the virus and spreading it via water in the Congo*

In the same context, Figure 7.9 shows a message spreading a rumor that the disease was sprayed from helicopters.

▶ Véranda
Mutsanga en Révolution
1 hour ago · 🌐

URGENT :VIRUS EBOLA .

A ce moment pendant vette nuit ;les
hélicoptères survolent le territoire de lubéro à
basse altitude pour injecté le virus à la
population de la dite territoire .

Figure 7.9. *Tweet from a Congolese man spreading a rumor that the disease was sprayed from helicopters. For a color version of this figure, see www.iste.co.uk/sidhom/systems.zip*

7.3.2. Regarding the experts

With social networks, "experts" can now exchange among themselves, set up actions, collective reflections and form networks of expertise on these media (Alloing and Moinet 2010).

In this context, health organizations, experts and research organizations have also taken over the social web to communicate, raise awareness and inform populations. Many studies have shown that in this new environment, Twitter is now used by many public health organizations to inform, educate and monitor the health of populations, especially in the event of a disaster (Hart et al. 2017).

In our study, it was mainly the national and international health organizations responsible for the response to the disease who spoke on

Twitter during the study period. In particular, we observed that they shared many tweets to inform the general public. For example, the Ministry of Health of the DRC made a regular update on the disease (see Figure 7.10), which it relayed on Twitter.

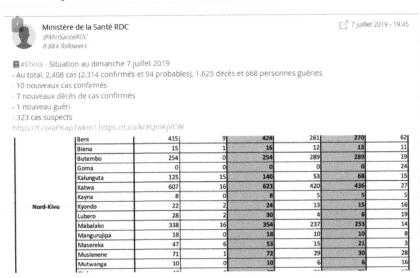

Figure 7.10. *Tweet relaying a daily update on the disease from the Congolese Ministry of Health. For a color version of this figure, see www.iste.co.uk/sidhom/systems.zip*

7.3.3. *Regarding the media*

Medical topics, including societal health debates, health disasters and global epidemics, have always been topics of interest to the media. Health issues thus undergo a double transformation in the media space: quantitative in their degree of social visibility and qualitative in their politicization (Gerstle and Piar 2016). At a time when the challenges of health and medicine weigh heavily on the choices of society, the press is thus called upon by all those who intervene in this debate, in our study, public and political men, experts, organizations of health, the citizen and more generally society.

Mathien insists on the visibility of health topics in the media. For him

to note that health has become a particularly privileged sector in the fields of observation and investigation of the mainstream media, means that it has left, in many aspects, professional and specialized circles (Mathien 1999, translated).

It is also the observation of Champagne who

observed, during the 1980s and 1990s, a strong development of medical information in the general information media, in the medical press and in the specialized press (Champagne 1999, translated).

The theme of health presents two types of space: a public societal space and a specialized space, and therefore offers the opportunity to study their confrontation (Zappala 1997). The first is largely driven by the media and the general press; the second is driven by specialized media specific to the scientific community (Zappala 1997). In this context, the advent of Web 2.0 has made it possible to open up these borders.

In the context of our study, the media are the vehicle for informing the debate in the democratized digital public space. They are also the vector of communication between the actors of the public debate. They are also involved. They can give their opinions and are likely to guide or even catalyze discussions. In any event, they are part of the debate.

In our work, it is mainly journalists and the press, both Congolese and international, who speak out. They tend to share WHO response releases, and prevention and awareness messages (see Figure 7.11).

Figure 7.11. *Tweet from the media relaying a preventive message to mobilize to fight the Ebola epidemic. For a color version of this figure, see www.iste.co.uk/sidhom/systems.zip*

7.3.4. *Regarding the politicians*

Public figures and political parties have been communicating via the Internet for over a decade. The way they have used this tool and the uses they make of it have already given rise to a certain number of works (Greffet 2001; Sauger 2002; Villalba 2003).

Their communication on the Web is done in particular via social networks such as Twitter, websites or blogs of political parties, as well as via personal websites or blogs, which may or may not be hosted by the party concerned (Longhi 2013).

They also use more conventional media such as the local, regional or national daily press, or react via blog posts associated with these online media (Gerstle and Piar 2016).

They are also communicated via institutional, administrative and legislative documents (parliamentary, administrative reports, etc.). According to Poupard, it is "an activity located at the crossroads of editorial, technical and socio-organizational issues" (Poupard 2005, translated).

Social networks have also enabled public and political figures to find a considerable strike force which has extended their area of influence (Maarek 2007), in particular to increase the opportunities to reach the population (Barboni and Treille 2010).

In our work, it is mainly Congolese politicians who speak out. They relay in particular prevention, health education or awareness messages. For example, Figure 7.6 shows a tweet relaying the photo of the President of the Congolese Republic (H.E.M. Felix Tshisekedi), who complies with medical requirements during his national tours (see Figure 7.12).

In conclusion, our study thus found 12 main influencers and highlighted a negative message tone of 73.31%.

Figure 7.12. *Screenshot of a tweet showing the president of the DRC applying hygienic gestures during his national tour. For a color version of this figure, see www.iste.co.uk/sidhom/systems.zip*

7.4. Conclusion

The Internet is seen as a considerable source of health information. Social networks were a source of health information with particularly high added value. Social networks make it possible to follow epidemics, especially the rumors and the actors communicating around these rumors (Tanti and Alate 2020).

The 2014–2015 Ebola epidemic was an interesting source of analysis of these rumors that have been the subject of a number of publications (Jin et al. 2014; Fung et al. 2016).

Our work focused on the 2018–2019 epidemic, and shows a certain number of interesting results. For example, unlike the results obtained during the 2014–2015 epidemic, our study highlights a shared Congolese population. Some people accept the disease and adhere to the treatment, and another party views it as a conspiracy (Tanti and Alate 2020).

Our study suffers from several limits. The short analysis period, like the choice to limit the study to the French language, is questionable. In addition, in our analysis, we did not take into account other media, such as Facebook and forums. Finally, it is especially the choice of analysis on the social network Twitter itself which is questionable. Indeed, this media limits the number of characters present in messages. It thus limits long discussions, therefore making it the relay of current events and the engine of polemics and debates, rather than the federator of true micro-communities.

In general, as a recommendation to change the mind of the defiant population, it would seem interesting to involve the relays of traditional healers and religious leaders who are the first to be consulted by this population in the communication campaign on social networks.

7.5. Acknowledgment

The author would like to thank Dr Alate Kodzo who participated in this project.

7.6. References

Alloing, C. and Moinet, N. (2010). Des réseaux d'experts à l'expertise 2.0. Le web 2.0 modifie-t-il la création et la mise en place de réseaux d'experts ? *Les Cahiers du numérique*, 1(6), 35–53.

Barboni, T. and Treille, É. (2010). L'engagement 2.0. Les nouveaux liens militants au sein de l'e-parti socialiste. *Revue française de science politique*, 60(6), 1137–1157.

Boyadjian, J. (2014). Twitter, un nouveau "baromètre de l'opinion publique" ? *Participations*, 8, 55–74 [Online]. Available at: https://doi.org/10.3917/parti. 008.0055.

Breton, P. (2000). *Le culte de l'Internet : une menace pour le lien social ?* La Découverte, Paris.

Champagne, P. (1999). Les transformations du journalisme scientifique et médical. In *Médias, santé, politique*, Mathien, M. (ed.). L'Harmattan Communication, Paris.

Colloc, J. (2014). Aspects éthiques et impact social du Big Data en santé. *Journées Big Data Mininf & Visualisation*, Journées EG C et AFIHM.

Ebola Outbreak Epidemiology Team (2018). Outbreak of Ebola virus disease in the Democratic Republic of the Congo, April–May, 2018: An epidemiological study. *Lancet*, 392(10143), 213–221.

Fung, I.C.-H., Fu, K.-W., Chan, C.-H., Chan, B.S.B., Cheung, C.-N., Abraham, T., Tse, Z.T.H. (2016). Social media's initial reaction to information and misinformation on Ebola, August 2014: Facts and rumors. *Public Health Reports*, 131(3), 461–473 [Online]. Available at: https://doi.org/10.1177/003335491613100312.

Gerstle, J. and Piar, C. (2016). *La communication politique*, 3rd edition. Armand Colin, Paris.

Greffet, F. (2001). Les partis politiques français sur le web. In *Les partis politiques : quelles perspectives ?* Andolfatto, N., Dominique, G., Fabienne, O.L. (eds). L'Harmattan, Paris.

Hart, M., (2017). Twitter and public health (Part 2): Qualitative analysis of how individual health professionals outside organizations use microblogging to promote and disseminate health-related information. *JMIR Public Health and Surveillance*, 3, e54 [Online]. Available at: https://doi.org/10.2196/publichealth.6796.

Jin, F., Wang, W., Zhao, L., Dougherty, E., Cao, Y., Lu, C.-T., Ramakrishnan, N. (2014). Misinformation propagation in the age of Twitter. *Computer*, 47(12), 90–94.

Longhi, J. (2013). Essai de caractérisation du tweet politique. *L'information grammaticale*, 136, 25–32.

Maarek, P. (2007). *Communication politique et marketing de l'homme politique*. Lexis Nexis, Paris.

Mathien, M. (1999). La santé dans la quête du bonheur dans la cité. In *Médias, santé, politique*, Mathien, M. (ed.). L'Harmattan Communication, Paris.

Medley, A.M. (2020). Case definitions used during the first 6 months of the 10th Ebola virus disease outbreak in the Democratic Republic of the Congo – Four neighboring countries, August 2018–February 2019. *Morbidity and Mortality Weekly Report*, 69(1), 14–19.

Poupard, J. (2005). Écrits d'écran : du mélange des genres. *Communications et langages*, 144, 65–76.

Quentin, J. and Citton, Y. (2015). Manifeste pour des humanités numériques 2.0. *Multitudes*, 2(59), 181–195.

Rieffel, R. (2014). *Evolution numérique, révolution culturelle ?* Gallimard, Paris.

Rodota, S. (1999). *La démocratie électronique. De nouveaux concepts et expériences politiques.* Apogée, Rennes.

Sauger, N. (2002). Les partis sur le Net : première approche des pratiques virtuelles des partis politiques français. In *L'Internet en politique, des États-Unis à l'Europe*, Serfaty, V. (ed.). Presses Universitaires de Strasbourg.

Tanti, M. (2012). Pandémie grippale 2009 dans les armées : l'expérience du veilleur. *Médecine & Armées*, 40(5), 389–401.

Tanti, M. and Alate, K. (2020). Quelle utilité des réseaux sociaux pour l'analyse des épidémies ? L'exemple de Tweeter pour le suivi de l'épidémie d'Ebola 2018–2019 en Afrique. *OCTA International Multi-Conference on Organization of Knowledge and Advanced Technologies*, Tunisia, 6–8 February.

Villalba, B. (2003). Moving towards an evolution in political mediation? French political parties and the new ICTs. In *Political Parties and the Internet, Net Again?* Gibson, R., Nixon, P., Ward, S. (eds). Routledge, London.

Zappala, A. (1997). La médecine médiatisée : entre la médicalisation du social et la socialisation de la science. *Hermès*, 21, 181–190.

From Human and Social Indexing to Automatic Indexing in the Era of Big Data and Open Data

8.1. Introduction

Today, information occupies a central place in our daily life. It represents a source of knowledge and power. In the era of Big Data and Open Data, a huge amount of information, documents, multimedia content and social tags are created, managed and stored electronically, which explains the exponential growth of data flows from a wide variety of fields that have led to the creation of an unprecedented amount of data. With this huge amount of data, it is becoming increasingly difficult to respond to user queries that are looking for relevant document results (Khemiri and Sidhom 2020). This is why new methods and algorithms have emerged to better represent the information collected from heterogeneous sources. In order to make these documents usable, a human (i.e. manual or intellectual) and/or an automatic indexing process allows us to create a document representation by a list of metadata, descriptors and social tags. These representations are used to find relevant information in a scalable collection of documents, to respond to user requests (information needs). In this context, numerous research works have been carried out to put forward indexing approaches. The ultimate goal of these different approaches is to better represent content (documents, electronic content, Big Data and Open Data) to effectively identify those that are most relevant when searching for information. This chapter presents a

Chapter written by Nabil KHEMIRI and Sahbi SIDHOM.

state of the art of approaches and methodologies ranging from manual and automatic indexing to algorithmic methods in the era of Big Data and Open Data.

8.2. Indexing definition

Indexing is a process of representing information which consists of identifying the significant elements to characterize multimedia documents (i.e. audio, images, text, video). This process analyzes documents to assign or extract a list of descriptors, metadata and social tags. These representations will subsequently facilitate the search for information in a collection of documents. With this in mind, an information retrieval system (IRS) must be composed of three principal functions (see Figure 1):

– representing document content;

– representing user information needs (user request or query);

– comparing these two representations in order to find documents, ordering the search results by relevance and returning the documents to the user.

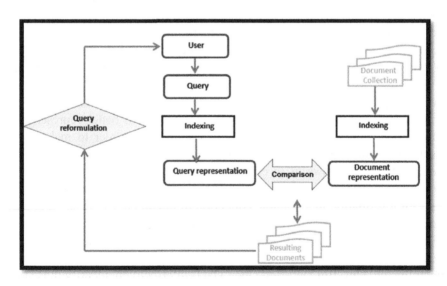

Figure 8.1. *Information retrieval system. For a color version of this figure, see www.iste.co.uk/sidhom/systems.zip*

Consequently, the performance of the IRS depends on the choice of the representation model and the matching process. There are three approaches that can be distinguished: the manual indexing approach, the automatic indexing approach and the semi-automatic indexing approach (combining an automatic approach with a manual approach).

8.3. Manual indexing

Manual indexing (i.e. intellectual or human indexing) is based on associations between words in a document with controlled vocabulary terms (manually assigned indexing terms). The choice of terms that represent each document (descriptors) depends on the know-how of the indexer, their knowledge and practical experience in the field of indexing. The human indexer uses a documentary language such as the thesaurus which provides a hierarchical dictionary (controlled vocabulary) of pre-established monolingual or multilingual standard terminologies to index documents. This type of approach allows classification and research by concepts (subjects or themes) in a collection of documents. Manual indexing is the result of document content analysis which is based on the following four steps:

– Documentary analysis (document analysis): the indexer must have global knowledge of the document to be analyzed. To analyze a document, they first consult the title, the table of contents, the summary, the introduction, the introductions and conclusions of the chapters (if they exist) and the conclusion. This speed-reading allows the indexer to identify the main subject or theme discussed or described in the document.

– The choice of concepts (keywords): to define the main concepts that best characterize a document, the indexer must answer a certain number of questions, those that a user would ask when searching for information such as: who and what is the document about? Where and when?

– Conversion of concepts into descriptors: the indexer chooses the appropriate index terms (the descriptors) from a controlled vocabulary list. A controlled vocabulary is a finite set of index terms from which all index terms must be selected. Only the approved terms can be used by the indexer to describe the document which ensures uniformity in the representation of the document (Chaumier and Dejean 1990).

– Proofreading and revision: during this step, the indexer decides to retain or reject some descriptors.

Human indexing has several disadvantages. It is costly in terms of money, vocabulary building time and assignment of concepts (index terms) to documents. It is subjective, since the choice of indexing terms depends on the indexer and their level of knowledge of the target domain. Although indexers follow the same steps, different concepts can be selected to characterize the same document. Also, controlled documentary language is difficult to maintain since the terminology is constantly evolving. When there is a high volume of documents, manual indexing becomes tedious and practically inapplicable (Clavel et al. 1993). Given the limits of manual indexing, the time and performance requirements, some documentary functions such as manual indexing should be automated.

8.4. Automatic indexing

With the advent of computers, researchers have realized that they can use automatic techniques and software methods to index a collection of documents in order to facilitate searching for information and obtain precise results with a reduced time and resources. Several factors have encouraged computer scientists, library and information science researchers to find new automatic methods that try to enrich or replace manual indexing. The automation of indexing has helped to overcome the limits and inadequacies of intellectual indexing approaches such as cost and subjectivity. Unlike human indexing, automatic indexing uses a free vocabulary formed by extracting key terms (a single word or a group of words) characterizing documents. Many statistical and/or linguistic indexing methods have been proposed to automatically extract the representative terms of a document.

8.4.1. *Statistical indexing methods*

Statistical methods of automatic indexing are based on purely mathematical and statistical calculations in order to define the weight of a word, according to different criteria such as:

– Word frequency: the weight of a word is calculated according to the number of occurrences (how many times a word appears in a document).

The most frequent words in the document are the most significant and will serve as a descriptor. We can eliminate unimportant words (i.e. stop words, grammatical words). Stop words are basically a set of commonly used words in any language. Determinants, pronouns, prepositions, conjunctions and grammatical adverbs are stop words. We should rather focus on the important words (i.e. content words, open-class words and lexical words) those that have meaning. Nouns, adjectives, verbs and adverbs are content words;

– Word density: the density of a word is calculated according to the ratio between its occurrence in the document and the size of this document.

– Word position in a document: the word position in a document can have an influence on its weighting. For example, the position of the word in the title is more advantageous than at the end of the document.

– Word writing style: give the advantage to words in capital letters and in bold in the weighting.

– Etc.

In information retrieval, there is a multitude of similarity measures in the literature. The best-known are TF-IDF (Term Frequency-Inverse Document Frequency) (Salton and McGill 1983), Dice similarity (Sneath and Sokal 1973), Jaccard similarity (Grefenstette 1994), character n-gram similarity (Shannon 1984), hidden Markov models (Baum and Petrie 1966), Levenshtein distance (Levenshtein 1965) and the Jaro–Winkler measure (Jaro 1989).

8.4.2. *Linguistic indexing methods*

Linguistic methods of automatic indexing are a subdomain of natural language processing (NLP). NLP is a multidisciplinary field that combines linguistics, computer science, information engineering and artificial intelligence. These methods use different levels of analysis:

– Morphological analysis is made up of three steps:

  - 1) Segmenting (tokenization) the text into sentences: a sentence is a string of characters located between a capital letter and a strong punctuation mark: full stop (period or full point), question mark and exclamation mark. The full stop as a sentence separator can present ambiguities. It can be an abbreviation marker or titles prefixing the name of a person (e.g. Mr, Mrs, Mrs, Dr, etc.), part of an acronym (e.g. I.S.K.O.).

– 2) Segmenting sentences into words: a word is a single distinct meaningful element of speech or writing, used with others (or sometimes alone) to form a sentence. The separators are spaces, numbers and weak punctuation marks (usually: comma, semicolon, colon, parentheses, ellipsis (also called suspension points), dash, brackets and quotation marks).

– 3) Lexical analysis is composed of lexical and inflectional morphological analysis: lexical morphological analysis consists of studying the form of words which can be simple, complex (compounds), variable (nouns, verbs, determiners, pronouns and qualifying adjectives) or invariable (adverbs, prepositions and coordinating conjunctions). Inflectional morphological analysis consists of studying the variation of lexical units as a function of grammatical factors. It represents the relationship between the different parts of a sentence and can concern a verb (conjugation) or a nominal group which depends on its grammatical category, and whether it is singular or plural.

– Syntactic analysis (or parsing) allows us to highlight the syntactic structure of a sentence by explaining the dependency relationships between words. The purpose of this phase is to represent the structure of sentences using syntax trees. Syntactic analysis identifies syntactic groups such as noun phrases, verb phrases, prepositional phrase, etc. These phrase groups are the basis of several indexing approaches (Chevallet and Haddad 2001; SIDHOM 2002; Bellila Heddaji 2005).

8.4.3. *Semantic indexing*

The problem of indexing documents by words, or groups of words, is not using semantic relationships between descriptors such as synonymy, homonymy, polysemy relationships, etc. With the emergence of terminological resources such as ontologies (Jonquet et al. 2010), semantics has become a major challenge to consider. Semantic indexing (Hamadi 2014; Yengui 2016) uses the concepts and their relationships to represent documents and queries.

8.4.4. *Social indexing*

Social indexing is a Web 2.0 technology, also known as social tagging, collaborative tagging, collaborative indexing social classification and Folksonomy. It involves a community of users freely creating and managing personalized tags (a keyword or term) assigned to a web resource for the purposes of collaborative categorization and classification. "Users are also actively involved in content creation, feedback and enrichment" (Rückemann 2012). Social indexing can create a shared content collaborative enrichment web and creating new communities. There are several research studies on social indexing such as recommendations in social networks (Dahimene 2014; Beldjoudi et al. 2016; Jelassi et al. 2016), improving information retrieval (Badache 2016), information monitoring (Pirolli 2011), etc.

8.5. Indexing methods for Big Data and Open Data

The rise of Big Data (or massive data, huge data) has followed the evolution of data storage and processing systems, notably with the advent of the technology of cloud computing (virtualization) and supercomputers. Big Data is also data but of a huge size. Big Data is a term used to describe a heterogeneous data set that is huge in volume and yet growing exponentially with time. These data sets are *so* voluminous and complex that none of the traditional data management tools are able to store and manage them efficiently. Doug Laney (Laney 2001) uses three properties or dimensions to define Big Data usually called the 3 Vs of Big Data (volume, variety and velocity). Volume refers to the growing volume of data generated through social media, websites, portals, online applications and connected objects (smart objects). Variety refers to the many types of data that are available which can be structured, semi-structured or unstructured, such as multimedia documents. Multimedia documents require additional preprocessing and a classification of the incoming data into various categories. Velocity refers to the speed with which data is being generated, received, stored, processed, analyzed and exploited in real time. Big Data comes from various sources, such as published content on Internet, messages exchanged on social media, data transmitted by connected objects, climate data, demographic data, scientific and medical data, data from sensors, e-commerce transactions, company data, etc.

Open Data is an important source of data; it refers to digital data whose access, use, re-use and redistribution (sharing) are public and free of rights (there should be no discrimination against persons or groups). They can be from the public or private sector, produced and published by the government, a public service, a community or by a company. The operation of this data offers numerous opportunities and new perspectives to improve the performance of companies and to extend human knowledge in many fields.

The huge volume of data, the variety of structures and types of multimedia documents from heterogeneous sources are the biggest problems when it comes to indexing. To overcome these problems, all indexed documents must be stored in the same format. The NoSQL (Not Only Structured Query Language) databases (Moniruzzaman and Hossain 2013; Bathla et al. 2018) are flexible and increasingly used with the rise of Big data to improve the performance of processing and analysis of distributed data. This data can have variable data structures different from those used by default in traditional relational databases. NoSQL databases do not use the rows/columns/table format. The most common types of NoSQL databases are key-value, wide column, document and graph:

– Key-value store: a key-value database, or key-value store, is a data storage paradigm designed for storing data in unique key-value pairs where each key is associated only with one value in a collection.

– Wide-column store: wide-column databases are designed for storing data as sections of columns where each key is associated only with a set of columns.

– Document store: document databases use common notation formats like JavaScript object notation (JSON) or extensible markup language (XML) to store documents. Each key is associated with a collection of key-value pairs stored in documents. This type of database is used to store structured and semi-structured documents.

– Graph store: the graph database uses graph theory to model data with nodes (entities or objects) and relationships (edges). Both nodes and relationships can have properties. This database type can store and analyze complex, dynamic and interconnected data. Many emerging problems such as social network analysis, network routing, trend prediction, product recommendation and fraud detection can be represented using graph models and solved using graph algorithms (Skhiri and Jouili 2013).

8.6. Conclusion

Indexing is a process used to extract descriptive elements from documents and user requests. The aim of indexing is to improve searching for information by finding relevant documents in a collection of documents in a reduced search time (Khemiri and Sidhom 2020). Several studies have been developed to suggest indexing approaches and methodologies ranging from manual and automatic methods to the emerging indexing methods for Big Data. This variety of methods must adapt and take advantage of continuous technological evolution. In recent years, the emergence of Big Data, Open Data and NoSQL databases have opened up a new technological era and new research areas. The purpose of these indexing methods is to allow the exploitation of huge digital data produced daily by humans and connected objects.

8.7. References

Bathla, G., Rani, R., Aggarwal, H. (2018). Comparative study of NoSQL databases for big data storage. *International Journal of Engineering & Technology*, 7. DOI: 10.14419/ijet.v7i2.6.10072.

Baum, L.E. and Petrie, T. (1966). Statistical inference for probabilistic functions of finite state Markov chains. *Annals of Mathematical Statistics*, 37(6), 1554–1563. DOI: 10.1214/aoms/1177699147.

Beldjoudi, S., Seridi, H., Benzine, A. (2016). Améliorer la recommandation de ressources dans les folksonomies par l'utilisation de linked open data. *IC2016: Ingénierie des Connaissances*, Montpellier, June.

Bellia Heddadji, Z. (2005). Modélisation et classification de textes : application aux plaintes liées à des situations de pollution de l'air intérieur. Doctoral thesis, Université Paris Descartes.

Chaumier, J. and Dejean, M. (1990). L'indexation documentaire, de l'analyse conceptuelle à l'analyse morphosyntaxique. *Documentaliste*, 27(6), 275–279.

Chevallet, J.-P. and Haddad, H. (2001). Proposition d'un modèle relationnel d'indexation syntagmatique : mise en œuvre dans le système iota. *INFORSID 2001*, Genève-Martigny.

Clavel, G., Walther, F., Walther, J. (1993). Indexation automatique de fonds bibliothéconomie. *ARBIDO-R8*, 14–19.

Dahimene, M.R. (2014). Filtrage et recommandation sur les réseaux sociaux. Doctoral thesis, École Doctorale Informatique, Télécommunication et Électronique, Paris.

Grefenstette, G. (1994). *Exploration in Automatic Thesaurus Discovery*. Kluwer Academic Publishers, London.

Hamadi, A. (2014). Utilisation de contexte pour l'indexation sémantique des images et vidéos. Thesis, Université Joseph Fourrier Grenoble.

Ismail, B. (2016). Recherche d'information sociale : exploitation des signaux sociaux pour améliorer la recherche d'information. Université Paul Sabatier – Toulouse III.

Jaro, M.A. (1998). Advances in record-linkage methodology as applied to matching the 1985 census of Tampa, Florida. *Journal of the American Statistical Association*, 89, 414–420.

Jelassi, M.N., Benyahia, S., Engelbert, M.N. (2016). Étude du profil utilisateur pour la recommandation dans les folksonomies. *IC2016 : Ingénierie des Connaissances*, Montpellier, June.

Jonquet, C., Coulet, A., Shah, N., Musen, M. (2010). Indexation et intégration de ressources textuelles à l'aide d'ontologies : application au domaine biomédical. *21èmes Journées Francophones d'Ingénierie des Connaissances*, Bordeaux.

Khemiri, N. and Sidhom, S. (2002). De l'indexation intellectuelle à l'indexation automatique à l'ère des Big Data et Open Data : un état de l'art. *OCTA Multiconference Proceedings: International Society for Knowledge Organization (ISKO-Maghreb)*, 170, February, Tunis [Online]. Available at: https://multiconference-octa.loria.fr/multiconference-program/.

Laney, D. (2001). 3D data management: Controlling data volume, velocity and variety. META Group Research Note, 6.

Levenshtein, V.I. (1965). Binary codes capable of correcting deletions, insertions, and reversals. *Doklady Akademii Nauk SSSR*, 163(4), 845–848.

Moniruzzaman, A.B.M. and Hossain, S. (2013). NoSQL database: New era of databases for big data analytics – Classification, characteristics and comparison. *International Journal of Database Theory and Application*, 6.

Pirolli, F. (2011). Pratiques d'indexation sociale et démarches de veille informationnelle. *Études de communication*, 53–66. DOI: 10.4000/edc.2615.

Rückemann, C.P. (2012). Integrated information and computing systems for natural, spatial, and social sciences. *Information Science Reference*, 543.

Salton, G. and McGill, M.J. (1983). *Introduction to Modern Information Retrieval.* McGraw-Hill Book Co., New York.

Shannon, C.E.A. (1984). Mathematical theory of communication. *Bell System Technical Journal*, 27(3), 379–423.

Sidhom, S. (2002). Plateforme d'analyse morpho-syntaxique pour l'indexation automatique et la recherche d'information : de l'écrit vers la gestion des connaissances. Doctoral Thesis, Université Claude Bernard, Lyon.

Skhiri, S. and Jouili, S. (2013). Large graph MINING: Recent developments, challenges and potential solutions. *Business Intelligence*, 103–124.

Sneath, P.H. and Sokal, R.R. (1973). *Numerical Taxonomy: The Principles and Practice of Numerical Classification.* W.H. Freeman and Company, San Francisco.

Yengui, A. (2016). Système de recherche d'information sémantique pour les bases de visioconférences médicales à travers les graphes conceptuels. Research paper, Faculté des sciences économiques et de gestion Sfax, Tunisia.

Strategies for the Sustainable Use of Digital Technology by the AWI in the Management of Knowledge and Cultural Communication on the "Arab World"

9.1. Introduction

Today more than ever, in the post-digital era, the management, production and transmission of content face major and multiple challenges. The stakes become all the greater when the content that is seen, read and listened to, conditions knowledge and scholarship around a civilization or geographical area. Such is the case with the Arab World Institute (AWI). In the same way, knowledge management and cultural communication have a special place in sustainable development, which has become a concept of primary interest in global policies for several years (CGLU 2015). The optimization of cultural tools has now become a necessity, and even an emergency. This chapter lies in this context, as it is part of a several-year doctoral work focused on the study of the images of the "Arab World" produced by the AWI via its public activities (Abassi 2020). It is also based on a recent work that analyzes the digital communication strategy of the AWI and its digital media. We also based our study on the work of M. Zacklad, L. Collet, F. Paquienseguy, E. Saïd, S. Faucheux, C. Hue, I. Nicolaï, F. Flipo, F. Deltou, M. Dobré, P. Lévy, H. Jenkins, G. Paquette and many other authors, which were of great help to us in the development of this research.

Chapter written by Asma ABBASSI.

In this chapter, which lies at the crossroads of digital humanities, cultural studies and information, communication sciences and sustainable development, we suggest studying the relationships between the material and immaterial worlds, the management methods of knowledge through the various digital tools, the images conveyed around the "Arab World" via the construction, organization and the transmission of knowledge from the AWI. We also study the feedback issue and the degree of interaction of audiences, mainly those in the "Arab World". We also discuss digital sustainability and durable knowledge management.

First of all, we will present the AWI by questioning its monopoly on knowledge around the Arab World in France and in the West in general. In a second step, we will analyze the differentiated uses of digital tools by the AWI in its communication strategy, based in particular on cross-media. Then, we will bring out the traces of a transmedia approach in which this cultural structure begins to engage. Thirdly, we will give an overview of the images built by the AWI and deal with the question of the feedback issue. Finally, we will tackle the question of the role of digital tools in sustainability and durability in the management of knowledge and cultural communication at the AWI.

9.2. The Arab World Institute and the construction of knowledge around the "Arab World" in the West

The idea of creating an institution dedicated to the "Arab World" in France was initiated in 1973 by Valéry Giscard d'Estaing. However, it was not until 14 years later that the AWI opened its doors to the public. Its opening dates back to November 30, 1987 by François Mitterrand, the French President at the time.

The AWI is a cultural foundation under French law created in 1987 in Paris, following a partnership between France and the 22 countries of the Arab League, which are, today, Algeria, Saudi Arabia, Bahrain, Comoros, Djibouti, Egypt, United Arab Emirates, Iraq, Jordan, Kuwait, Lebanon, Libya, Morocco, Mauritania, the Sultanate of Oman, Palestine, Qatar, Somalia, Sudan, Syria, Tunisia and Yemen. This institution was founded in a postcolonial context, at a time in history marked by defiance towards the Muslim Arab world, when thinkers advocated the end of the Great Stories (Lyotard 1979). The AWI would come to translate the secular links between

France and the "Arab World", the crossed influences between East and West and the strong and tumultuous historical relationships where the ideological aspect has a great place. Geopolitical, economic, security and cultural aspects are not to be outdone either.

Since its creation, the AWI has depicted itself as the "mirror" of the "Arab World" and its culture in France and the West. With the advent of digital means, this unique structure has changed its system of construction and transmission of knowledge with differentiated uses of digital tools, alternating between documentary and communication strategies.

Indeed, this a priori cultural foundation depicts itself as a place of exchange and enhancement of Arab culture and civilization, where the West can be "at the gates of the Arab World" and view the history, the present and the future of this part of the planet. The AWI presents itself as a "bridge" between the East and the West, as the exclusive holder of knowledge and information about the "Arab World", this conceptual and strategic area, not only in France, but also in Europe. Furthermore, when it was founded, it announced that "The Institute is mainly a showcase for the public; thanks to which France would become the capital of modern Arab culture." This would also increase the stakes of the AWI, since it attracts an international audience and enjoys great visibility: we are talking about a million visitors a year. Consequently, the representation that it conveys has a tremendous effect. Actually, it would reflect, through a game of mirrors, so many visions of this part of the world and its people.

Thus, we wonder if the AWI really plays the role it pretends to play, and if it is truly representative of the "Arab World". It is necessary to question the nature of the image or images that this composite and multicultural entity reflects, especially that it has a weight not least in the world scene. Hence, the importance of the challenges of knowledge organization (KO) that we consider is far beyond the meaning of its document indexing.

Throughout our research (Abassi 2017), we have come to the conclusion that the AWI is undoubtedly France's diplomatic and political tool. This foundation has survived so far despite budget deficits, scandals and problems of all kinds; it is precisely because of the political stakes that hide more or less well behind its cultural activity. It is an important tool in the French diplomacy of influence on the great territory of the "Arab World" and even in the world. Through its museum and public activities, which are mainly exhibitions, meetings and debates, cinema, music and dance, the AWI offers Western audiences a fragmentary, very selective, sometimes insidious and reducing image that is far from reflecting the reality and current events of the "Arab World" and its people. Most of the time, it corresponds to a post-orientalist imagination and phantasmagoria.

However, it should be noted that the image of the Arab World is less important than the way it is constructed. Indeed, in the post-digital era, this cultural structure has appropriated digital tools, social networks in particular. It now offers to the French and international audience an image and representations of this great territory that is the "Arab World" through the scheduling of its public activities, as well as through the content that it broadcasts on the digital tools, namely an institutional website and digital social networks: Facebook (2010), Twitter (2012), Instagram (2014), SoundCloud and YouTube (2008). There are so many interfaces that make up its "digital showcase".

The AWI has had an institutional website since 1997. The website has since undergone several changes both in terms of its architecture and graphic design, and its data organization and content. The information and digital communication policy mainly began in 2013, a year which coincides with the appointment of Jack Lang as head of the AWI, a leading French political figure who was, among others, the Minister of Cultural affairs with François Mitterrand when the foundation came into being. He is a politician and a man of the arts who gives great importance to communication. We should note that "digital projects" dealing with the Internet, social networks and audiovisual products were an integral part of the AWI's activity reports over the past decade, hence their importance and strategic aspect in the eyes of the decision-makers. In 2012, the AWI also recruited Yannis Koikas as director of digital mediation, which has become a priority issue.

9.3. The AWI's digital communication strategies

According to activity reports from 2013 to 2015, several audits were carried out for the website, which has adopted new editorial and publication charters; it has reorganized its content and its ergonomics, graphics and access to information via a new revised interface: a complete remake was carried out. A web mastering policy and a social media strategy have also been implemented, with the drafting of an editorial charter, the launching of educational publications related to current news with rather moderate comments. The AWI has adopted a policy of audiovisual recording of its various events and has produced trailers and presentation films for the exhibitions that it publishes on its YouTube channel, Facebook, and even Twitter in certain cases, as well as on SoundCloud, in audio mode of course. Web documentaries and interviews with artists and personalities of Arab or French origin have also been produced.

There is no doubt that the AWI has engaged in "all digital" in recent years and has emphasized digital mediation with the various online resources and services, whatever their form or medium. The AWI has finally "gone digital", concretizing the will of its president Jack Lang who would like "to position the AWI as a leading cultural player on the Web".

As mentioned above, the AWI, which has taken the digital "turning point", provides Internet users with information-documentation and information-communication, while also focusing on branding. This information, which is textual, iconic, videographic and audio, is shared on the various current digital tools in a separate way. This differentiated use is conditioned by the nature of the interface, like SoundCloud, which only supports audio tracks or YouTube, which is intended for audiovisual documents with some textual elements. But there is also a question of strategy and content orientation, among other things, depending on the nature of the network and the profile of its subscribers. The AWI has opted for the multiplication of information "relays" as well as their formats, and the specialization and targeting of content to increase its visibility and strengthen public loyalty. Its digital information offer is closely related to its structure and its public activities that actually take place. Here, the intangible world only exists to better value the material one: the intangible ecosystem is closely linked to the physical universe, but knowledge and scholarship are organized in a fragmentary way because information, namely digital resources, is dispatched and structured in an uneven and different way from one support/interface to another. The AWI is developing an approach to segment content and therefore knowledge and scholarship, in order to reach its target audience everywhere. This logic can be qualified as proactive. It produces an enriched cultural content that enters digital humanities conditioning – or not – the knowledge and understanding of the "Arab

World" by a Western and Arab audience. Through this policy, the AWI aims for a sustained, continuous digital media visibility.

According to Woorank analysis, the AWI's institutional website is the 395,002th most visited site in the world. It is also perfectly optimized for smartphones and touch screens, with a classification and taxonomic system and multiple entries. Moreover, it is the most exhaustive. It is available in three languages and contains a database concerning the AWI as a structure (activities, practical information, organization chart and operation, e-shop, etc.) and documentary resources whose purpose is to "make you discover" the "Arab World" (its art, culture, science, history, language and scriptures, religion and society), as well as the museum and the digitization of its collections. Thus, we can say that we are talking about a digitization of heritage and a non-immersive guided tour for the purpose of popularization and democratization and patrimonialization. We also find links to the various webdocs, websites and applications of the AWI, as well as to its publications and some of its bibliographies.

The website also provides an opportunity to find out about AWI's cultural programming, and current and past events. We should point out that information-communication becomes information-documentation as soon as it becomes obsolete, thus changing status. There is no doubt that the website, which also contains links to all of the institution's social networks, presents the most complete knowledge construction, organization and management around the Arab World, always through the prism of the AWI.

As for the Facebook page with its nearly 160,000 followers, 1,772 shares, 140 comments and 621 "likes" (according to Worank), it is used to communicate about events that take place at the AWI, with a frequency of around three posts a day; these posts contain a call text (or presentation), an image or a video and a link to the website (for "Reservation" or "More info"), mainly concerning exhibitions, shows and news, at the center of which Jacques Lang is often present. These are mainly trails for promoting activities, teasers and video topics for communication. The latter may also refer to events which take place outside the walls of the AWI, usually relating to artists of Arab origin ("The AWI recommends"), which is a novelty that indicates a certain openness to extramuros and the Franco-Arab artistic scene.

In total, the Facebook page contains 3,865 photos and 347 videos linked to the YouTube channel, as well as a number of event alerts. Internet users have the opportunity to comment, like and share posts. However, this type of interaction is not very important, although the activity reports of the AWI mention that they are clearly increasing.

Figure 9.1. *AWI's Facebook page (screenshot). For a color version of this figure, see www.iste.co.uk/sidhom/systems.zip*

The Instagram account of the AWI with its 32.1k subscribers has 654 publications in the form of images (an intrinsic characteristic of this social network), often annotated by a small presentation text. These images can be posters of exhibitions, works of art, portraits of artists, photographs taken from photographic or other exhibitions or covers of events of different types (photo report). On Instagram, the frequency of posts is much lower than on Facebook. The AWI does not present the same content at all on its Facebook and Instagram accounts, even if the two social networks are used to communicate about its events and news. Internet users therefore do not find the same content in an instant. On Instagram, we find a more developed interaction, especially at the likes level; however, some comments also exist.

Figure 9.2. *AWI's Instagram account (screenshot). For a color version of this figure, see www.iste.co.uk/sidhom/systems.zip*

Just like on the Facebook page, on the AWI Twitter account with its 34.9k subscribers, we find text, a link to the website and an image or video. This network gives Internet users the possibility to like, retweet, comment and share. Its content is different from that on Facebook, but sometimes has redundant information. We have noted a prevalence given to information concerning meetings and debates, as well as to exhibitions and news. The number of tweets can range from two to five per day.

For the YouTube channel, with its 14.7k subscribers and 4,069,778 views, it shows a film presenting the activities of the AWI on the home page. Then, the videos are organized by activity-theme as follows: exhibitions at the AWI with promotional films or teasers, portraits of artists, economic meetings of the "Arab World", meetings and debates, trailers, music at the AWI, educational actions, AWI museum and cinema.

There is also another entry through the "Playlist" and "Videos". These are recordings of meetings and debates, portraits of artists, etc. Each video is put online with its respective title and a short presentation text. The frequency of publications varies between two and four per month, but the number can be higher if there is more activity, and can reach 12 or even more videos per month.

Figure 9.3. *AWI's Twitter account (screenshot). For a color version of this figure, see www.iste.co.uk/sidhom/systems.zip*

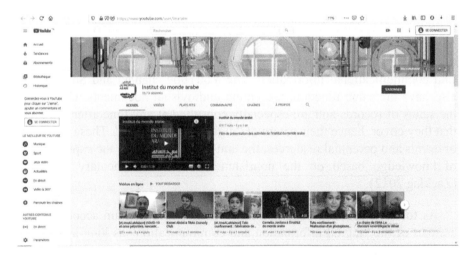

Figure 9.4. *AWI's YouTube channel (screenshot). For a color version of this figure, see www.iste.co.uk/sidhom/systems.zip*

However, for the YouTube channel, comments are disactivated, which makes any interaction with Internet users impossible. The latter can only like, dislike or share. Thus, we suspect that YouTube is only used to host videos because it is practical. In our opinion, this hypothesis is valid, since the same videos are called to leave Twitter and Facebook. This is an amazing use of YouTube, an exclusively AWI documentary base with an almost zero economic cost.

On SoundCloud, the "sound YouTube", with 535 followers, the AWI broadcasts audio content from "Thursdays of the AWI", "Literary meetings", "An hour with…". Currently, the public can listen to 341 tracks. The frequency of audio posts is less regular than on any other social network. For instance, the last publication dates back to six months, a period in which 16 audio tracks were put online in a month. The audio tracks occur at a random frequency from one month to another.

Accordingly, we can state that the AWI has taken the digital turn in a world where 50% of the people are connected, especially in Europe and North America (Lévy 2019). Also, the AWI builds knowledge and scholarship around it and the Arab World through cross-media and a differentiated use of digital tools and "distributed communication transactions" (Zacklad 2004), adapting both to the channels and their audiences. The AWI does distinguish between content and support, which conditions the modes of apprehension by the public (Zacklad 2012). Indeed, the website tends to favor the documentary aspect, even if within it we do find communicational information. The YouTube channel and SoundCloud also have the same function, presenting audiovisual documents which have the status of records and are especially consulted at any time after the event that they cover, hence their nature of "digital documentary". These are traces of events and perennial resources, the aim of which may be the capitalization of knowledge based on the nourishment of a "documentary memory" (Zacklag 2012).

As for the Facebook page and the Twitter and Instagram accounts, they are part of the "digital information and communication" as long as they are in the news and in the immediate future, increasing the proximity of the AWI with its connected audiences and thus participating in its visibility and that of its various activities. In fact, any communication about a particular event takes place in three stages: before, during and after. The objective is mercantile and promotional: to attract the public, build loyalty and encourage the act of purchasing – in terms of marketing – but there is also the development of fan culture, as well as branding and corporate communication, thus creating a digital identity (Abassi 2020).

Figure 9.5. *IM's SoundCloud account (screenshot). For a color version of this figure, see www.iste.co.uk/sidhom/systems.zip*

Regarding the digital use which favors tree and hypertext architectures (Collet and Paquienseguy 2015), the AWI uses many more tools than we have mentioned. The adventure and discovery video game "Medelia, the treasure of the Mediterranean", the application "Kalila Wa Dimna", the web doc "Did you say Arabic?", the web series "The submerged mysteries of Egypt" and the immersive and interactive site "Hip hop from the Bronx to the Arabs" were created to extend the experience and the visit of certain exhibitions, and Oculus Rift glasses with headphones were available to visitors of the "Contemporary Morocco" exhibition for an immersive experience with augmented reality. The content of the six setup cameras created a 360° video. Thus, we can say that the AWI has, in a certain way, entered documentary transmediality (Zacklad 2012), even if certain aspects are still lacking, such as the collective creation of content. The participative aspect is not very developed either.

However, we can say that the AWI offers some multi-entry transmedia narration from around the Arab World, even if it is partial. For example, digital tools can be understood independently, as well as in their globalness, they are meant for commercial and academic purposes, the images of the "Arab World" are available on several platforms; there is sometimes seriality, subjectivity, multiplicity and a construction of a universe with a certain aesthetic. If we quote Jenkins (Jenkins 2003; Jenkins et al. 2013), we can qualify the narrative universe presented as "fictional reality", as we will explain. This multichannel and multimedia access to knowledge and information, whether documentary or communicative, has cultural, economic, social and political challenges. We are talking about the issues of production, dissemination and reception of content and therefore knowledge and scholarship. What images of the Arab World does the AWI present through its construction and dissemination of knowledge? (Abassi 2020).

9.4. The images built by the AWI and the question of feedback

There is no doubt that through its digital communication strategy, now institutionalized and mobilizing different channels, the AWI presents a significant panel of Arab heritage. There is also an effort to reach out to the Other, the Arab, and a will – at least apparent – to discover it, as well as a certain dynamism of the Institute and of the "Arab World" in its consensual sense, an image which was not existent enough in the past. In fact, we witness notable changes in the discourse developed on the "Arab World" by the AWI, which recognizes certain realities of this geographical and cultural area. There is a real inscription in contemporaneity, through the treatment of current topics and through the presentation of young artists. Even though the foundation is still out of step with all of the new emerging artistic and cultural sensitivities in the "Arab World", its territorial roots. But what contemporaneity is it? It is about the French contemporaneity of the "Arab World". Indeed, since its foundation, the AWI favors Western artists and elite or Arabs residing in the West, in particular in France, sometimes with European origins. These are those who have been integrated into otherness, as the author Alain Mascarou pointed out so well in his book "Orientalism in its Mirrors". The number of Arab artists and elite residing in their countries of origin could be considered insignificant. Thus, the "standard bearers" of the "Arab World" are, in reality, only personalities with a double genealogy, an "offshore" elite, from immigrant backgrounds or not. This is likely to reduce the representativeness of the artistic and cultural scenes of the Arab countries and bias them. There is precisely a problem of immersion in the Arab reality and an ambiguity of an "in-between" glance concept. As we predicted in our thesis, the AWI tends to be a Franco-French structure of Arab culture intended for the French and the Arab community in France and in the West. The AWI is not the mirror of Arab culture in France as much as it is the mirror of French culture and representations of the "Arab World" in France.

In fact, France is at the top of the list of virtual visitors to AWI platforms and interfaces according to an analysis by Google Trend. France is trying to contain the scope of this prestigious showcase which is the AWI and use it as a containment tool. These are, indeed, the values of secularity, interbreeding and multiculturalism that this structure wants to spread (Abassi 2017). We also note, as we said above, a change in the positive meaning, in the discourse on the "Arab World"; more objective and more embedded in reality, even if a large part of this reality is completely omitted. Exoticism, post-orientalism and patrimonialization still have a great place. A heritage which is the fruit of the patrimonial policy which marked the French cultural policy that reached its peak with Jack Lang. Through its programming, the AWI conveys the history of the Arabs and their countries, most often focusing on their glorious times and the golden age of their civilization. The "Arab World" is generally condemned in its past. It is often presented in a biased manner in its contemporaneity. The mass public is thus comforted with a picturesque and postcolonial ritual. It is a fictional reality that the AWI creates and displays, always developing a discourse of legitimization generally based on celebration and glorification. But one of the problems of the AWI is that it is built on an entity, the "Arab World", which does not exist, a problem that officials are becoming aware of. The images of the "Arab World" constructed through public activities and the digital tools of the AWI are made from a few fragments of reality, which prevents the discovery of the Other, the Arab, or reach its real actuality, its truth. These images are meant for distribution to the general public, for mass consumption. This policy is indeed integrated into the economy of public culture; as culture is being incontestably considered as a true economic activity.

Another important point is that of feedback (Paquette 1987). Aimed at an audience that is resident in France, made up of Westerners of ethnic origin and/or of French people with Arab roots, feedback from the Arab World concerning this Western vision offered by the AWI is almost non-existent. Programming and media coverage and keeping records through digital devices are not intended for Arabs in their countries of origin, except for research and touristic purposes, such as visits to France. The level of interactivity in the AWI's social networks is generally not very developed. It therefore tends towards zero with regard to the Arab World; the latter has no power or control over the production and dissemination of the knowledge and scholarship that concerns it at the AWI. In the best of cases, an Arab will be able to comment on a Facebook post, like it or share it, and visit, in an immersive way, an in situ exhibition, at best. It is their only field of intervention and possibility. The participative dimension is non-existent, and interactivity is limited by the very nature of the digital tools, the possibility of content regulation is also impossible. The Arab user is positioned only as a passive receiver of knowledge and scholarship emitted by the AWI. Their role is therefore limited to reception-observation, which can reduce their appropriation of content, and this obviously does not fit in the stated objectives of the foundation.

9.5. The role of digital tools in sustainability and durability in the management of knowledge and communication at the AWI

As we have just seen, the AWI strategically adopts digital tools via differentiated uses and a transversal method combining cross-media and transmedia to communicate about its activities and the "Arab World"; it also communicates around a certain heritage and culture. Thus, the AWI registered itself through a cultural approach in durability and sustainability, on many aspects. Indeed, the managers of this foundation are optimizing the use of digital tools for the dissemination of culture which is the fourth pillar of a sustainable development (CGLU 2015), as ICTs are the keys (Faucheux et al. 2010). It goes without saying that the socio-digital networks with their very important information offer make it possible to control and reduce the costs of knowledge management and communication by offering ROI (Return of Investment) calculation tools and a large concentration of resources. Even if the AWI still uses traditional communication (urban posters and other print media), it decreases physical resources and mainly focuses its mediation with its audiences on new technologies, "body extensions", which have far-reaching effects and are less harmful to the environment, even if some people speak of "rebound effects" and refute the theory of green digitalization (Flipo et al. 2016). This digital infrastructure and network thus offers the possibility of replacing – even partly – the conventional means of communication and information with all the advantages it presents. This also allows the AWI to control the costs of its non-stop communication. Also, digital communication saves travel. In addition to saving time, it also limits the use of environmentally harmful means of transport. The knowledge building and cultural communication strategy adopted by the AWI in recent years therefore has a significant ecological potential. Connected Internet users benefit from a timely information offer that allows them to travel only to discover an exhibition, a show or take part in a debate or other event or activity in situ. Non-residents in France can follow the institution's activities from afar and discover certain aspects of the "Arab World" through feedback, various reports and resources posted online. There is no doubt that this dematerialization of cultural information and communication has many economic and social advantages (Petit 2009). Borders are blurring, creating a new digital geography offering another type of proximity (Ullmann et al. 2008). The culture of the "Arab World" through the prism of the AWI is now diffused in an immaterial and free-way on a multiplicity of interfaces, thus participating in the development of the concept of equity in access to information, even if a digital gap still exists around the world, creating excluded audiences. The reproducible nature of digital data makes it possible, in the same sense, to "get rid of the laws of scarcity, and to optimize all activities" (Laget 2008).

In addition to this, the digital tools adopted by the AWI contribute to durability. Knowledge and representations of the "Arab World" are stored and transmitted in time and space. They continuously extend and feed the images of this great geographic and conceptual territory constructed by the AWI. Thanks to new technologies, these images can be viewed as an unchanging "Arab World", lasting over time even if the nature and content of the images is changing. The design of this area and its identity, namely Arabity, through the activities of the foundation becomes perennial and sustainable. The digitization of data and the dematerialization of communication and procedures for the construction, management and transmission of knowledge allow a simple and rapid access to information (IRFEDD 2017), simplify the transmission codes and increase "sharability" (Laget 2008), even if the information is intensive. Documentary resources and communication thus become accessible to the greatest number of audiences, which is capable of ensuring a certain democratization. The relationships with time and space are changing, and distance is no longer an obstacle to accessing information (Papa et al. 2006). The knowledge and skills around the "Arab World" created by the AWI are now compiled, organized, updated, juxtaposed, referenced, available and accessible (IRFEDD 2017); consequently, audience targeting and performance of the communication strategy are optimized. Facility and speed of use are increased because digital technology, as we have already said, erases distances. Resources, which are intrinsically essential, are also optimized and valued in an intangible manner. Via digital means, their management becomes easier and under continuous construction. This contributes to popularization, even partially, and to the promotion of cultural heritage, the identity of a people, the creative potential of its members, diversity, creativity, etc., all contributing to sustainable development, including culture, which is a main component.

9.6. Conclusion

Digital tools have become the essential ally of the AWI in its digital cultural communication and in the dissemination of information and documentary resources (Abassi 2020). As we have seen, the foundation has recently undertaken new methods of access to information and knowledge around its activity and around the "Arab World". It has used new digital information devices, a plurality of technologies and multimedia writing for cultural mediation, thus conveying and sharing representations and knowledge divided into autonomous units. The information and documents provided by the AWI serve both to communicate and capitalize on knowledge of various kinds. Proposed in various forms and supports (several interfaces with several inputs), digital information is redundant and complementary (Collet and Paquienseguy 2015) and participates in a cross-media strategy, which at times tends towards transmedia dimensions. It is a semiotic production shared in a segmented and differentiated way. However, the digital circulation of knowledge remains one-sided, most of the time without feedback from the Arab people residing in their countries. The organization of knowledge is a reflection of the activities of the AWI, but it does not represent a true reflection of the reality of the "Arab World", the definition of which is still unclear and controversial. Obviously, the AWI, this structure which enjoys great visibility, has not grasped every historic moment in its complexity, did not reveal the "Arab World" with its various components and the contradiction of its realities. The AWI remains fatally frozen, circulating in a vacuum, restricted to countries, "representatives" and limited cultural forms, and enclosed in improved stereotypes which do not offer all the keys to the understanding of the contemporary "Arab World".

AWI officials have a personalized view of the "Arab World" which nourishes the imagination of their audiences from the knowledge of a subjective nature, especially as the notion of Arabity is more or less fallacious. We believe that the AWI offers representations and constructs images of the "Arab World" which contrast, in many respects, with the realities of this entity, and gives a version where the share of fiction and fabrication is important. This is due to considerations and issues that are essentially of a political and geostrategic order. Even though the displayed intentions are noble, this structure builds an "Arab World" fabricated and even prefabricated with scattered elements of reality. The resulting image is ideological and overloaded, biased and not representative of the reality of this great territory that is the "Arab World". Finally, we are dealing with an off-ground durability based on a dematerialized sustainability.

9.7. References

Abassi, A. (2017). Le "Monde arabe" de l'Institut du Monde arabe. PhD Dissertation in Sciences and Techniques of Arts, Higher Institute of Fine Arts of Tunis.

Abassi, A. (2020). Digital knowledge management and cultural communication: The Arab world according to the Arab World Institute (IMA). *International Multi-Conference on Organization of Knowledge And Advanced Technologies* (OCTA), IEEE, Tunisia.

Assemblée générale des Nations Unies (2015). Transformer notre monde : le Programme de développement durable à l'horizon 2030. UN, A/RES/70 [Online]. Available at: https://unctad.org/meetings/fr/SessionalDocuments/ares70d1_fr.pdf.

Bachimont, B. (1994). *Le contrôle dans les systèmes à base de connaissances : contribution à l'épistémologie de l'intelligence artificielle*. Hermès, Paris.

Barthes, R. (1957). *Mythologies*. Le Seuil, Paris.

Bonnafous, S. and Pailliart, I. (1994). Les territoires de la communication. *Mots*, 40, 134–136.

Bourdaa, M. (2013). Le transmedia storytelling. *Terminal*, 112 [Online]. Available at: http://journals.openedition.org/terminal/447.

CGLU (2015). Culture 21 : Actions. Engagements sur le rôle de la culture dans les villes durables. *Cités et Gouvernements Locaux Unis*, Barcelona [Online]. Available at: https://www.arts-ville.org/wp-content/uploads/2017/10/Culture21Actions_2015.pdf.

Collet, L. and Paquienseguy, F. (2015). Imaginaires, techniques et pratiques de l'écriture multimédia interactive : des cd-rom aux e-albums culturels. *Interfaces numériques*, 2(2), 13 [Online]. Available at: https://www.researchgate.net/publication/295134113_Imaginaires_techniques_et_pratiques_de_l'ecriture_multimedia_interactive_des_cd-rom_aux_e-albums_culturels.

European Alliance for Culture and the Arts (2017). No sustainable development without culture [Online]. Available at: https://allianceforculture.com/no-sustainable-development-without-culture/.

Faucheux, S., Hue, C., Nicolaï, I. (2010). *TIC et développement durable : les conditions du succès*. De Boeck, Brussels.

Flipo, F., Deltou, F., Dobré, M. (2016). Les technologies de l'information à l'épreuve du développement durable. *Natures Sciences Sociétés*, 24, 36–47. NSS-Dialogues, EDP Sciences [Online]. Available at: http://www.nss-journal.org.

Jenkins, H. (2003). Transmedia storytelling. Moving characters from books to films to video games can make them stronger and more compelling. *Technological Review* [Online]. Available at: www.technologyreview.com/news/401760/transmedia-storytelling/.

Jenkins, H., Green, J., Ford, S. (2013). *Spreadable Media: Creating Value*. NYU Press, New York.

Laget, M. (2008). Le numérique, simple mue du libéralisme ou avènement d'une économie soutenable. *Netcom*, 22–3/4 [Online]. Available at: http://journals.openedition.org/netcom/1618.

Lévy, P. (2019). Le rôle des humanités numériques dans le nouvel espace politique [Online]. Available at: http://www.sens-public.org/article1369.html?lang=fr.

Lyotard, J.-F. (1979). *La Condition postmoderne : rapport sur le savoir*. Collection Critique, Paris.

Numérique & Développement Durable (2017). Travaux du Conseil scientifique de l'IRFEDD [Online]. Available at: https://www.irfedd.fr/wp-content/uploads/2017/11/Publication-Conseil-scientifique-Num%C3%A9rique-et-d%C3%A9veloppement-durable.pdf.

Papa, F., Collet, L., Landel, P. (2006). TIC et développement durable : du maillage à la valorisation des territoires de montagne. *Revue de géographie alpine*, 94(3), 89–97 [Online]. Available at: http://www.persee.fr/web/revues/home/prescript/article/rga_0035-1121_2006_num_94_3_2409.

Paquette, G. (1987). Feedback, rétroaction, rétroinformation, réponse... du pareil au même. *Communication et langages*, 73, 5–18 [Online]. Available at: http://www.persee.fr/doc/colan_0336-1500_1987_num_73_1_984.

Petit, M. (2009). Les technologies de l'information et de la communication (TIC) au service du développement durable. *Annales des Mines – réalités industrielles*, 2, 83–88 [Online]. Available at: https://www.cairn.info/revue-realites-industrielles1-2009-2-page-83.htm.

Poirrier, P. (2000). *L'État et la culture en France au XXe sicècle*. Le Livre de Poche, Paris.

Saïd, E. (2005). *L'orientalisme. L'orient créé par l'Occident*. Le Seuil, Paris.

Ullmann, C., Vidal, P., Bourcier, A. (2008). L'avènement d'une société de l'information durable. *Netcom*, 22–3/4 [Online]. Available at: http://journals.openedition.org/netcom/1749.

Zacklad, M. (2004). Processus de documentation dans les documents pour l'Action (DopA) : statut des annotations et technologi es de la coopération associées. *Le numérique : impact sur le cycle de vie du document pour une analyse interdisciplinaire*, 13–15 October. Editions de l'ENSSIB & Montreal, Quebec.

Zacklad, M. (2012). Organisation et architecture des connaissances dans un contexte de transmédia documentaire : les enjeux de la pervasivité. *Études de Communication*, 39 [Online]. Available at: http://journals.openedition.org/edc/4017. DOI: 10.4000/edc.4017.

Zacklad, M. (2018). Nouvelles tendances en organisation des connaissances. *Études de Communication*, 50 [Online]. Available at: http://journals.openedition.org/edc/7566.

List of Authors

Asma ABBASSI
Institut Supérieur des Beaux-arts
University of Tunis
Tunisia

Pedro Chávez BARRIOS
ERPI Laboratory
University of Lorraine
Nancy
France
and
Autonomous University of
Querétaro
Santiago de Querétaro
Mexico

Inès BAYOUDH SÂADI
RIADI Laboratory
ENSI University of Manouba
Tunisia

Walid BAYOUNES
RIADI Laboratory
ENSI University of Manouba
Tunisia

Sabrine BEN ABDRABBAH
LARODEC Laboratory
ISG University of Tunis
Tunisia

Nahla BEN AMOR
LARODEC Laboratory
ISG University of Tunis
Tunisia

Hénda BEN GHÉZALA
RIADI Laboratory
ENSI University of Manouba
Tunisia

Malika BEN KHALIFA
LGI2A Laboratory
University of Artois
Béthune
France
and
LARODEC Laboratory
ISG University of Tunis
Tunisia

Mariem BRIKI
LARODEC Laboratory
ISG University of Tunis
Tunisia

Zied ELOUEDI
LARODEC Laboratory
ISG University of Tunis
Tunisia

Hammou FADILI
Pôle Recherche &
Prospective/FMSH
CNAM
Paris
France

Amira KADDOUR
ENSTAB
University of Carthage
Tunisia

Nabil KHEMIRI
LORIA Laboratory
University of Lorraine
France
and
University of Jendouba
Tunisia

Marilou KORDAHI
Saint-Joseph University
Beirut
Lebanon
and
Paragraph Laboratory
University of Paris 8
France

Eric LEFÈVRE
LGI2A Laboratory
University of Artois
Béthune
France

Davy MONTICOLO
ERPI Laboratory
University of Lorraine
Nancy
France

Sahbi SIDHOM
LORIA & INRIA Grand-Est
Laboratories
University of Lorraine
Nancy
France

Marc TANTI
Center of Armed Forces
Epidemiology and Public Health
UMR1252
SESSTIM
Aix Marseille University
France

Index

Other titles from

in

Computer Engineering

2022

ZAIDOUN Ameur Salem
Computer Science Security: Concepts and Tools

2021

DELHAYE Jean-Loic
Inside the World of Computing: Technologies, Uses, Challenges

DUVAUT Patrick, DALLOZ Xavier, MENGA David, KOEHL François,
CHRIQUI Vidal, BRILL Joerg
*Internet of Augmented Me, I.AM: Empowering Innovation for a New
Sustainable Future*

HARDIN Thérèse, JAUME Mathieu, PESSAUX François,
VIGUIÉ DONZEAU-GOUGE Véronique
*Concepts and Semantics of Programming Languages 1: A Semantical
Approach with OCaml and Python*
*Concepts and Semantics of Programming Languages 2: Modular and
Object-oriented Constructs with OCaml, Python, C++, Ada and Java*

VENTRE Daniel
Artificial Intelligence, Cybersecurity and Cyber Defense

2019

BESBES Walid, DHOUIB Diala, WASSAN Niaz, MARREKCHI Emna
Solving Transport Problems: Towards Green Logistics

CLERC Maurice
Iterative Optimizers: Difficulty Measures and Benchmarks

GHLALA Riadh
Analytic SQL in SQL Server 2014/2016

TOUNSI Wiem
Cyber-Vigilance and Digital Trust: Cyber Security in the Era of Cloud Computing and IoT

2018

ANDRO Mathieu
Digital Libraries and Crowdsourcing
(Digital Tools and Uses Set – Volume 5)

ARNALDI Bruno, GUITTON Pascal, MOREAU Guillaume
Virtual Reality and Augmented Reality: Myths and Realities

BERTHIER Thierry, TEBOUL Bruno
From Digital Traces to Algorithmic Projections

CARDON Alain
Beyond Artificial Intelligence: From Human Consciousness to Artificial Consciousness

HOMAYOUNI S. Mahdi, FONTES Dalila B.M.M.
Metaheuristics for Maritime Operations
(Optimization Heuristics Set – Volume 1)

JEANSOULIN Robert
JavaScript and Open Data

PIVERT Olivier
NoSQL Data Models: Trends and Challenges
(Databases and Big Data Set – Volume 1)

SEDKAOUI Soraya
Data Analytics and Big Data

SALEH Imad, AMMI Mehdi, SZONIECKY Samuel
Challenges of the Internet of Things: Technology, Use, Ethics
(Digital Tools and Uses Set – Volume 7)

SZONIECKY Samuel
Ecosystems Knowledge: Modeling and Analysis Method for Information and Communication
(Digital Tools and Uses Set – Volume 6)

2017

BENMAMMAR Badr
Concurrent, Real-Time and Distributed Programming in Java

HÉLIODORE Frédéric, NAKIB Amir, ISMAIL Boussaad, OUCHRAA Salma, SCHMITT Laurent
Metaheuristics for Intelligent Electrical Networks
(Metaheuristics Set – Volume 10)

MA Haiping, SIMON Dan
Evolutionary Computation with Biogeography-based Optimization
(Metaheuristics Set – Volume 8)

PÉTROWSKI Alain, BEN-HAMIDA Sana
Evolutionary Algorithms
(Metaheuristics Set – Volume 9)

PAI G A Vijayalakshmi
Metaheuristics for Portfolio Optimization
(Metaheuristics Set – Volume 11)

2016

BLUM Christian, FESTA Paola
Metaheuristics for String Problems in Bio-informatics
(Metaheuristics Set – Volume 6)

DEROUSSI Laurent
Metaheuristics for Logistics
(Metaheuristics Set – Volume 4)

DHAENENS Clarisse and JOURDAN Laetitia
Metaheuristics for Big Data
(Metaheuristics Set – Volume 5)

LABADIE Nacima, PRINS Christian, PRODHON Caroline
Metaheuristics for Vehicle Routing Problems
(Metaheuristics Set – Volume 3)

LEROY Laure
Eyestrain Reduction in Stereoscopy

LUTTON Evelyne, PERROT Nathalie, TONDA Albert
Evolutionary Algorithms for Food Science and Technology
(Metaheuristics Set – Volume 7)

MAGOULÈS Frédéric, ZHAO Hai-Xiang
Data Mining and Machine Learning in Building Energy Analysis

RIGO Michel
Advanced Graph Theory and Combinatorics

2015

BARBIER Franck, RECOUSSINE Jean-Luc
COBOL Software Modernization: From Principles to Implementation with the BLU AGE® Method

CHEN Ken
Performance Evaluation by Simulation and Analysis with Applications to Computer Networks

2013

ROCHANGE Christine, UHRIG Sascha, SAINRAT Pascal
Time-Predictable Architectures

WAHBI Mohamed
Algorithms and Ordering Heuristics for Distributed Constraint Satisfaction Problems

ZELM Martin *et al.*
Enterprise Interoperability

2012

ARBOLEDA Hugo, ROYER Jean-Claude
Model-Driven and Software Product Line Engineering

BLANCHET Gérard, DUPOUY Bertrand
Computer Architecture

BOULANGER Jean-Louis
Industrial Use of Formal Methods: Formal Verification

BOULANGER Jean-Louis
Formal Method: Industrial Use from Model to the Code

CALVARY Gaëlle, DELOT Thierry, SÈDES Florence, TIGLI Jean-Yves
Computer Science and Ambient Intelligence

MAHOUT Vincent
Assembly Language Programming: ARM Cortex-M3 2.0: Organization, Innovation and Territory

MARLET Renaud
Program Specialization

SOTO Maria, SEVAUX Marc, ROSSI André, LAURENT Johann
Memory Allocation Problems in Embedded Systems: Optimization Methods

2011

BICHOT Charles-Edmond, SIARRY Patrick
Graph Partitioning

BOULANGER Jean-Louis
Static Analysis of Software: The Abstract Interpretation

CAFERRA Ricardo
Logic for Computer Science and Artificial Intelligence

HOMES Bernard
Fundamentals of Software Testing

KORDON Fabrice, HADDAD Serge, PAUTET Laurent, PETRUCCI Laure
Distributed Systems: Design and Algorithms

KORDON Fabrice, HADDAD Serge, PAUTET Laurent, PETRUCCI Laure
Models and Analysis in Distributed Systems

LORCA Xavier
Tree-based Graph Partitioning Constraint

TRUCHET Charlotte, ASSAYAG Gerard
Constraint Programming in Music

VICAT-BLANC PRIMET Pascale *et al.*
Computing Networks: From Cluster to Cloud Computing

2010

AUDIBERT Pierre
Mathematics for Informatics and Computer Science

BABAU Jean-Philippe *et al.*
Model Driven Engineering for Distributed Real-Time Embedded Systems

BOULANGER Jean-Louis
Safety of Computer Architectures

MONMARCHÉ Nicolas *et al.*
Artificial Ants

PANETTO Hervé, BOUDJLIDA Nacer
Interoperability for Enterprise Software and Applications 2010

SIGAUD Olivier *et al.*
Markov Decision Processes in Artificial Intelligence

JUSSIEN Narendra
A TO Z OF SUDOKU

2006

BABAU Jean-Philippe *et al.*
From MDD Concepts to Experiments and Illustrations – DRES 2006

HABRIAS Henri, FRAPPIER Marc
Software Specification Methods

MURAT Cecile, PASCHOS Vangelis Th
Probabilistic Combinatorial Optimization on Graphs

PANETTO Hervé, BOUDJLIDA Nacer
Interoperability for Enterprise Software and Applications 2006 / IFAC-IFIP I-ESA'2006

2005

GÉRARD Sébastien *et al.*
Model Driven Engineering for Distributed Real Time Embedded Systems

PANETTO Hervé
Interoperability of Enterprise Software and Applications 2005

Printed and bound by CPI Group (UK) Ltd, Croydon, CR0 4YY

27/10/2024

14580732-0001